鱼类疾病诊断和防治图谱

——细菌、病毒卷

Manual of Fish Diseases:
Diagnosis,Prevention and Management
(Bacterial and Viral Diseases)

夏新生　薛汉宗　主编
By Xinsheng Xia, Thomas Sit

中国农业出版社
CHINA AGRICULTURE PRESS
北　京
BEIJING

本书编委会
Editors and authors

主编 Chief editors	夏新生	Xinsheng Xia	薛汉宗	Thomas Sit
副主编 Associate editors	卢体康	Tikang Lu	何展豪	Kenny Ho
	屈 娟	Juan Qu	白诺文	Christopher Brackman
	秦智峰	Zhifeng Qin	耿 捷	Jie Geng
	张剑锐	Jianrui Zhang	刘 荭	Hong Liu
编写人员 Contributing authors	黄倩君	Qianjun Huang	何汉邦	Jeremy Ho
	贾 鹏	Peng Jia	纪 凡	Fan Ji
	霍颖彤	Eunice Fok	蔡慧珊	Jenny Tsoi
	何正贻	Abigail Ho	史秀杰	Xiujie Shi
	阮柏文	Carlton Yuen	兰文升	Wensheng Lan
	毕 丹	Denise Bi	郑晓聪	Xiaocong Zheng
	王津津	Jinjin Wang	温智清	Zhiqing Wen
	刘 莹	Ying Liu	于 力	Li Yu

前 言

由深圳海关和香港渔农自然护理署共同编撰的《鱼类疾病诊断和防治图谱——病毒、细菌卷》，在2017年开始搜集图片资料，并以中英双语撰写文字，并经过多次审核、校对。本书内容包括文字和图片，扼要介绍及描述疾病病原和其流行病学、诊断方法、防治方法、临床症状、剖检及组织病理等，覆盖鱼类细菌、病毒和立克次氏体等疾病。书籍内容简明扼要，适用于学校教学、水生动物疾病检测技术培训、兽医（水生动物类）临床诊断、出入境检疫人员口岸现场检疫、疫病监测和流行病学调查等。

对于部分病毒病病原名称，国际病毒分类委员会进行了变更，但是考虑到行业仍沿用旧称，本书遵从行业习惯，仍采用旧称，特此说明。

编　者

2020年12月

Foreword

The book, *Manual of Fish Diseases: Diagnosis, Prevention and Management* (*Bacterial and Viral Diseases*), written by Shenzhen Customs (SZC) and the Agriculture, Fisheries and Conservation Department (AFCD) of the Hong Kong Special Administrative Region, was given turns of reviewing and detailed editing. The development of this book started with the searching and collection of photographs in 2017, followed by subsequent writing of the textual content in both Chinese and English. This book includes textual content, photographs and figures concisely describing and introducing the pathogens covering fish bacterial, fungal and viral diseases and their respective epidemiology, diagnostics methods, prevention and control measures as well as clinical signs, post-mortem and histopathological changes. The book provides concise contents of fish diseases which would be suitable to be applied for purposes including but not limited to academic education, technical training of aquatic animal disease, disease diagnosis for aquatic veterinarians, on-site inspection for import and export quarantine, disease surveillance and epidemiological investigation.

The International Committee on Taxonomy of viruses has changed the term principle of some viral pathogens, however, the origin terms are still in use. Accordingly, in the current publication, we followed the occupation practice and still adopts the origin name. Hereby declaration.

Dec, 2020

目录 Contents

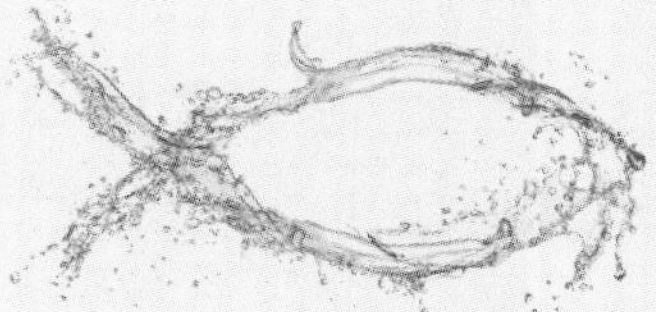

鲤春病毒血症

疾病概述

【概述】 鲤春病毒血症是一种严重的急性出血和流行性败血症。

【宿主】 易感宿主包括鲤（*Cyprinus carpio*）、锦鲤（*Cyprinus carpio*）、鲫（*Carassius carassius*）、鲢（*Hypophthalmichthys molitrix*）、鳙（*Aristichthys nobilis*）、草鱼（*Ctenopharyngodon idellus*）、金鱼（*Carassius auratus*）、高体雅罗鱼（*Leuciscus idus*）、丁鲹(*Tinca tinca*)、欧鲇（*Silurus glanis*）和虹鳟（*Oncorhynchus mykiss*）；潜在病毒携带者：蓝鳃太阳鱼（*Lepomis macrochirus*）、大口黑鲈（*Micropterus salmoides*）、东方蝾螈（*Cynops orientalis*）、欧鳊（*Abramis brama*）。

【易感阶段】 所有年龄阶段均可被感染，1龄以下的幼鱼最易出现临床症状并发病。

【发病水温】 发病温度11 ～ 17℃。在寒冷的国家，严冬过后鱼的生理状况较差，容易发病。

【地域分布】 2000年前仅分布于欧洲，后在巴西、中国、美国和加拿大有报道。

【疾病地位】 世界动物卫生组织（OIE）将其列入水生动物疫病名录。

病原

（1）2017年，国际病毒分类委员会将病原名称改为鲤春病毒（Carp sprivivirus）。过去病原名称为鲤春病毒血症病毒（Spring viraemia of carp virus，SVCV）。

（2）属弹状病毒科（*Rhabdoviridae*）、鲤春病毒属（*Sprivivirus*）。

（3）病毒呈弹状，长度为80 ～ 180nm，直径为60 ～ 90nm，有囊膜。

（4）病毒基因组为不分节段、负义、单股的RNA，包含有11 019个核苷酸；编码5个蛋白，分别是核蛋白（N）、磷蛋白（P）、基质蛋白（M）、糖蛋白（G）和聚合酶（L）。

（5）对编码糖蛋白基因550个核苷酸的片段进行基因分析，将SVCV与其相近病毒分为Ⅰ、Ⅱ、Ⅲ、Ⅳ 4个不同的基因型。SVCV都属于Ⅰ型。Ⅰ型又分为Ⅰa、Ⅰb、Ⅰc、Ⅰd 4个亚型。

临床症状和病理学变化

（1）病鱼嗜睡，脱离鱼群或聚集在水口或者池塘边，有些身体失去平衡。

（2）体色变暗，眼突出，鳃发白，鳍条基部和泄殖孔出血，泄殖孔突出，并拖有黏稠的粪便。

（3）鳃丝变性，腹胀或者水肿，肠道发炎，内脏器官水肿出血，肌肉、脂肪组织和鱼鳔有点状出血。

（4）肝脏和血管出现炎症并坏死。肝脏实质充血，有多个点状坏死和退变。心脏出现心包炎，呈点状退变和坏死。脾脏充血，网状内皮增殖，巨噬细胞中心增大。胰脏发炎，有多个坏死病灶。肾脏泌尿组织和造血组织出现损伤，肾小管堵塞，细胞透明化并出现空泡。肠道血管周边发炎，上皮脱落，微绒毛萎缩。腹膜发炎。鳔上皮层由单层变为不连续的多层，黏膜下血管因淋巴细胞浸润而扩张。

诊断方法

（1）病毒分离 使用鲤上皮瘤细胞系（EPC）、肥头鲤细胞系（FHM）和草鱼性腺细胞系（GCO），分离温度为20℃。

（2）半嵌套式RT-PCR 引物为SVCV-F1（5′-TCT-TGG-AGC-CAA-ATA-GCT-CAR-RTC-3′）和SVCV-R2（5′-AGA-TGG-TAT-GGA-CCC-CAA-TAC-ATH-ACN-CAY-3′）。退火温度为55℃，扩增产物长度为714bp。第二步引物为SVCV-F1（5′-TCT-TGG-AGC-CAA- ATA-GCT-CAR-RTC-3′）和SVCV-R4（5′-CTG-GGG-TTT-CCN-CCT-CAA-AGY-TGY-3′）。退火温度为55℃，扩增产物长度为606bp。扩增产物经测序后确定。

（3）酶联免疫吸附试验（ELISA）和间接免疫荧光试验（IFAT） Testline ELISA试剂盒兔源多抗的特异性不够高，Bio-X的IFAT试剂盒鼠源单抗特异性过强，因此，血清学诊断不能用于确诊。

防治方法

（1）保持养殖场良好的卫生管理水平。

（2）对苗种和观赏鱼实行严格检疫。

（3）清塘消毒，杀灭病原体。

（4）在冬季和早春季节降低养殖密度，可以减少病毒的扩散。

（5）内服免疫增强与代谢调节剂，如多糖、多肽和多种维生素等。

（6）将水温升高到20℃以上，可终止或有效防止SVC的暴发。

Spring viraemia of carp, SVC

Disease overview

[Disease Characteristic] Acute disease, causing severe haemorrhage and septicemia.

[Susceptible Host] Common carp (*Cyprinus carpio*), koi (*Cyprinus carpio*), crucian carp (*Carassius carassius*), silver carp (*Hypophthalmichthys molitrix*), big head carp (*Aristichthys nobilis*), grass carp (*Ctenopharyngodon idellus*), gold fish (*Carassius auratus*), Ide (*Leuciscus idus*), tench fish (*Tinca tinca*), wels catfish (*Silurus glanis*) and rainbow trout (*Oncorhynchus mykiss*). Potential carriers of Carp sprivivirus include bluegill (*Lepomis macrochirus*), largemouth bass (*Micropterus salmoides*), Chinese fire belly newt (*Cynops orientalis*) and common bream (*Abramis brama*).

[Susceptible Stage] Fish of any age can be affected. Clinical signs and disease occurrences are usually noted in juveniles below 1 year of age.

[Outbreak Water Temperature] Disease outbreaks occur at 11~17℃. In cold countries, it is mostly prevalent after winter when the physiological condition of the fish is compromised.

[Geographic Distribution] Mainly distributed in Europe before the year of 2000, afterwards, the disease was reported in the United States of America, Brazil, Canada and China.

[Disease Status] The World Organisation for Animal Health (OIE) Listed Aquatic Animal Disease.

Aetiological agent

(1) Spring viraemia of carp virus (SVCV). It was renamed as Carp sprivivirus by The International Committee on Taxonomy of Viruses (ICTV) in 2017.

(2) Family: *Rhabdoviridae*. Genus: *Sprivivirus*.

(3) Bullet-shaped, 80~180nm in length, 60~90nm in diameter, with capsule.

(4) Viral genome is a non-segmented, negative-sense, single-stranded RNA. Containing 11,019 nucleotides which encode 5 proteins, namely nucleoprotein (N), phosphoprotein (P), matrix protein (M), glycoprotein (G) and polymerase (L).

(5) Sequencing analysis of a 550 nucleotide region encoding the glycoprotein gene showed that SVCV and similar virus can be separated into four distinct genotypes as Ⅰ, Ⅱ, Ⅲ and Ⅳ. SVCV belongs to genotype Ⅰ, which can further be subdivided into four sub-types, namely Ⅰa, Ⅰb, Ⅰc and Ⅰd.

Clinical signs and pathological changes

(1) Lethargic, segregated or gathered at the edge or water inset of the pond, some may lose balance.

(2) Darkened body colour, exophthalmia, pale gills, haemorrhages at the base of fins and the vent, protruding vent with trailing mucoid faecal casts.

(3) Abdominal distension or oedema, degeneration of gill lamellae, enteritis, visceral oedema or haemorrhage, focal haemorrhage of muscles, adipose tissues and swim bladder.

(4) Inflammation and necrosis in the liver and blood vessels. Hyperaemic liver with degeneration and multifocal necrosis. Pericarditis with focal degeneration and necrosis. Hyperaemic spleen with hyperplasia of the reticuloendothelium and enlarged melanomacrophage centres. Pancreatitis with multiple necrotic foci. Damage to the excretory and haematopoietic systems of

the kidney, with cells undergoing vacuolation and hyaline degeneration blocking the renal tubules. Perivasculitis, desquamation of the epithelium and villi atrophy in the intestines. Peritonitis. Metaplasia of the epithelium of the swim bladder from single layer to multiple discontinuous layers, dilated vessels with lymphocytic infiltration in the sub-mucosa.

Diagnostic methods

(1) **Virus isolation** Use EPC, FHM and GCO cell-lines and incubate at 20℃ .

(2) **Semi-nested reversed transcriptase-polymerase chain reaction (RT-PCR)** Use the primers SVCV-F1 (5′-TCT-TGG-AGC-CAA-ATA-GCT-CAR-RTC-3′) and SVCV-R2 (5′-AGA-TGG-TAT-GGA-CCC-CAA-TAC-ATH-ACN-CAY-3′) with annealing temperature at 55℃ .The amplicon size is 714bp. Run semi-nested PCR on the amplicon with primers SVCV-F1 (5′-TCT-TGG-AGC-CAA-ATA-GCT-CAR-RTC-3′) and SVCV-R4 (5′-CTG-GGG-TTT-CCN-CCT-CAA-AGY-TGY-3′) with annealing temperature at 55℃. The amplicon size of the second PCR is 606bp. Perform sequencing on the amplicon for confirmation.

(3) **Enzyme-linked immunosorbent assay (ELISA) and Immunofluorescence antibody test (IFAT)** The specificity of the commercially available Testline ELISA kit is not high enough, while the specificity of the mouse source monoclonal antibodies in Bio-X IFAT test kit is too high. Therefore, serology test would not be applicable for confirmatory testing.

Preventative measures

(1) Maintain good hygiene management in aquaculture farms.

(2) Quarantine inspection of seedlings and ornamental fish.

(3) Stamp out the pond to eradicate the pathogen.

(4) Reduce farming population density in winter and early spring to minimize spreading of the virus.

(5) Oral administration of immunostimulant and metabolic modifiers such as polysaccharides and multi-vitamins, etc.

(6) Increase the water temperature to over 20℃ may effectively prevent the outbreak of SVC.

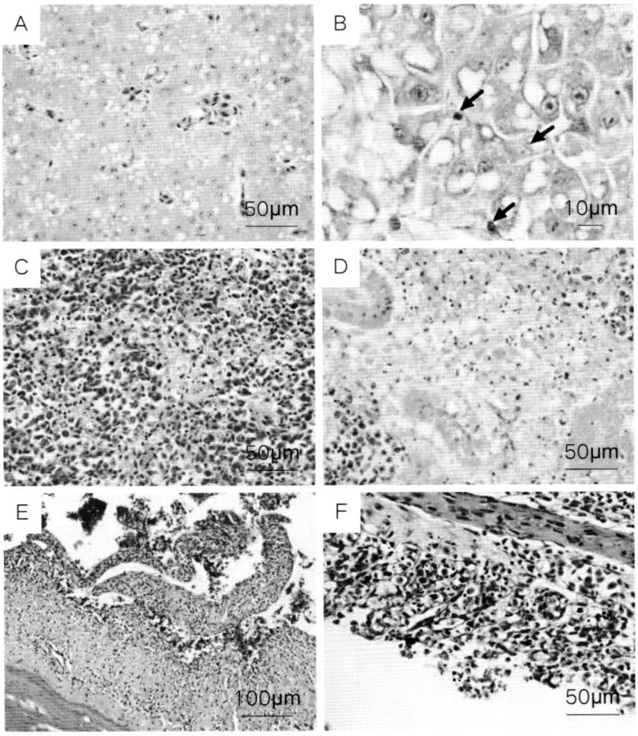

受感染的鲈鲤细胞及组织病变（HE染色）

A.肝脏充血和空泡变性

B.局灶性坏死伴肝细胞核固缩和核溶解（箭头）

C.脾脏充血出血，巨噬细胞和网状内皮增生

D.肾小管上皮变性坏死，造血组织坏死

E.伴有上皮坏死和脱落的卡他性肠炎

F.鳃坏死和出血

[源自ZHENG et al.，2018]

Pathologic lesions in infected *Percocypris pingi*（HE staining）

A. Hyperaemia and vacuolar degeneration in the liver

B. Focal necrosis with karyopyknosis and karyolysis (arrows) of liver cells

C. The spleen was hyperaemia and hemorrhage, with hyperplasia of macrophages and the reticuloendothelium

D. Degeneration and necrosis of the renal tubular epithelium, and necrosis of the haematopoietic tissue

E. Catarrhal enteritis with necrosis and sloughing of the epithelium

F. Necrosis and hemorrhage in the gill

[Source：ZHENG et al.，2018]

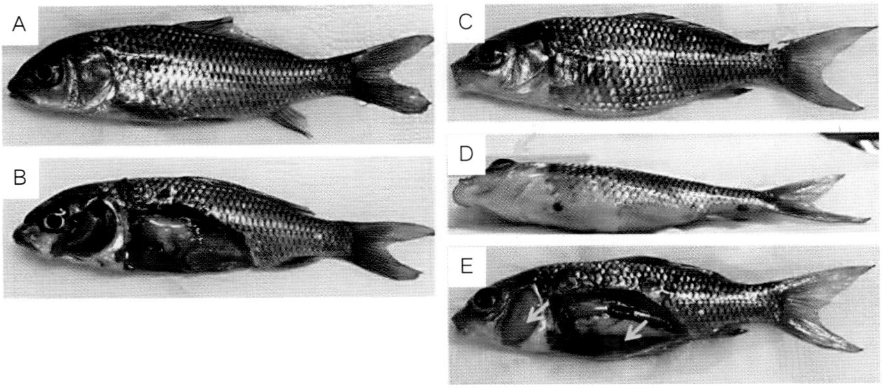

患病鲤

A ~ B为正常鲤，患病鲤皮肤出血（C）、眼球突出（D）、鳃发白，出血（箭头）（E）

[源自GODAHEWA et al.，2018]

Macroscopic findings of affected fish

A and B indicated the morphology of un-infected common carp. Infected fish showed widespread dermal hemorrhage (C), pop-eye (D), and pale gills, internal hemorrhages (E) are indicated by yellow arrows

[Source：GODAHEWA et al.，2018]

传染性造血器官坏死

疾病概述

【概述】 传染性造血器官坏死是一种高度传染性急性流行病。

【易感宿主】 主要感染鲑科鱼类。自然条件下，感染病原的种类有虹鳟（*Oncorhynchus mykiss*）、大鳞大麻哈鱼（*O. tshawytscha*）、红大麻哈鱼（*O. nerka*）、大麻哈鱼（*O. keta*）、玫瑰大麻哈鱼（*O. rhodurus*）、马苏大麻哈鱼（*O. masou*）、银大麻哈鱼（*O. kisutch*）、大西洋鲑（*Salmo salar*）、褐鳟（*S. trutta*）、克拉克大麻哈鱼（*O. clarki*）、湖红点鲑（*Salvelinus namaycush*）、北极红点鲑（*S. alpinus*）、美洲红点鲑（*S. fontinalis*）、远东红点鲑（*S. leucomaenis*）、香鱼（*Plecoglossus altivelis*）、欧洲鳗鲡（*Anguilla anguilla*）、太平洋鲱（*Clupea pallasi*）、大西洋鳕（*Gadus morhua*）、高首鲟（*Acipenser transmontanus*）、白斑狗鱼（*Esox lucius*）、海鲫（*Cymatogaster aggregata*）和管吻刺鱼（*Aulorhychus flavidus*）。

【易感阶段】 幼苗阶段是易感期，随着鱼龄增长，鱼群抗感染能力逐渐增强。有细菌混合感染、处于应激状态下，亚临床感染会变成显性感染。产卵期的鱼群高度易感，其排出的性腺产物中含有大量病毒。

【发病水温】 自然条件下，8 ～ 15℃发病。急性发病时，累积死亡率可达90%～ 95%。

【地域分布】 在北美洲（加拿大和美国）、亚洲（伊朗、日本、朝鲜和中国）、欧洲（奥地利、比利时、克罗地亚、捷克、法国、德国、意大利、荷兰、波兰、俄罗斯、斯洛文尼亚、西班牙、瑞士）均有分布。南半球尚未发现有分布。

【疾病地位】 世界动物卫生组织（OIE）将其列入水生动物疫病名录。

病原

（1）2017年，国际病毒分类委员会将病原名称改为鲑粒弹状病毒（Salmonid novirhabdovirus）。过去病原名称为传染性造血器官坏死病毒（Infectious haematopoietic necrosis，IHNV）。

（2）属弹状病毒科（*Rhabdoviridae*）、粒外弹状病毒属（*Novirhabdovirus*）。

（3）病毒呈弹状，长度为150 ～ 190nm，直径为65 ～ 75nm，有囊膜。

（4）病毒基因组为不分节段、负义、单股的RNA，包含有11 131个核苷酸，基因组序列中4种核苷酸G、A、T、C含量分别为24.28%、28.67%、19.46%、27.59%。编码6个蛋白，分别是核蛋白（N）、磷蛋白（P）、基质蛋白（M）、糖蛋白（G）、非结构蛋白（NV）和聚合酶（L）。

（5）对编码糖蛋白基因303个核苷酸的片段进行基因分析，将IHNV分为U、L、E、J和M 5个基因型。不同地区鱼群感染的IHNV属于不同的基因型，其中，从太平洋东北部的红大麻哈鱼分离到的IHNV为U基因型，美国加利福尼亚州的大鳞大麻哈鱼为L基因型，欧洲虹鳟为E基因型，亚洲虹鳟为J基因型，美国爱达荷州的虹鳟为M基因型。

临床症状和病理学变化

（1）病鱼昏睡，运动迟缓，有时狂暴乱窜。

（2）体表变暗，鳃苍白，眼球突出，有拖尾的排泄物。

（3）有腹水，腹部肿胀，体表有出血斑。

（4）剖检可见贫血，肠道缺乏食物，肝脏、肾脏和脾脏苍白，内脏器官有出血斑。

（5）造血组织、脾脏、肾脏、肝脏、胰腺和消化道出现变性、坏死，肠壁嗜酸性粒细胞坏死。

诊断方法

（1）病毒分离 使用鲤上皮瘤细胞系（EPC）、肥头鲤细胞系（FHM），分离温度为15℃。

（2）RT-PCR 引物为IHNV-F（5′-AGA-GAT-CCC-TAC-ACC-AGA-GAC-3′）和IHNV-R（5′-GGT-GGT-GTT-GTT-TCC-GTG-CAA-3′）。退火温度为50℃，扩增产物长度为693bp，扩增产物经酶切或测序后进一步诊断。

（3）酶联免疫吸附试验（ELISA）

防治方法

（1）保持养殖场良好的卫生管理水平。

（2）清塘消毒，杀灭病原体。

（3）内服免疫增强与代谢调节剂，如多糖、多肽、多种维生素等。

（4）注射商品化疫苗。

Infectious haematopoietic necrosis, IHN

Disease overview

[Disease Characteristic] IHN is a highly contagious and acute epidemic disease.

[Susceptible Host] Mostly affect salmonid species. In natural condition, the following species can be affected: rainbow trout (*Oncorhynchus mykiss*), chinook or king salmon (*O. tshawytscha*), sockeye or red salmon (*O. nerka*), biwa trout (*O. rhodurus*), cherry salmon *(O. masou)*, silver salmon *(O. kisutch)*, Atlantic salmon *(Salmo salar)*, brown trout *(S. trutta)*, cutthroat trout *(O. clarki)*, lake trout *(Salvelinus namaycush)*, Arctic charr *(S. alpinus)*, brook trout *(S. fontinalis)*, whitespotted charr *(S. leucomaenis)*, sweetfish *(Plecoglossus altivelis)*, European eel *(Anguilla anguilla)*, Pacific herring *(Clupea pallasi)*, Atlantic cod *(Gadus morhua)*, white sturgeon *(Acipenser transmontanus)*, northern pike *(Esox lucius)*, shiner perch *(Cymatogaster aggregate)* and tube-snout *(Aulorhychus flavidus)*.

[Susceptible Stage] Fry are the most susceptible stage while older fish is more resistant to the disease. Clinical disease may occur in fish with other sub-clinical infection, when under stress or with bacterial co-infection. During spawning period, fish is highly susceptible and may shed large amount of virus in the gonad secretions.

[Outbreak Water Temperature] In natural condition, disease outbreaks occur at 8~15℃. In acute cases, the cumulative mortality rate can reach 90%~95%.

[Geographic Distribution] Widely distributed in North America （United States of America，Canada)，Asia (Iran, Japan, Korea，China)，Europe (Austria, Belgium, Croatia, Czech Republic, France, Germany, Italy, Netherlands, Poland, Russia, Slovenia, Spain, Switzerland). No distribution has been found in the southern hemisphere.

[Disease Status] OIE-listed Aquatic Animal Disease.

Aetiological agent

(1) Infectious haematopoietic necrosis virus (IHNV), renamed as Salmonid novirhabdovirus by the International Committee on Taxonomy of Viruses (ICTV) in 2017.

(2) Family：*Rhadboviridae*. Genus: *Novirhabdovirus*.

(3) The virus is bullet-shaped, capsulated, 150~190nm in length and 65~75nm in diameter.

(4) Viral genome is non-segmented, negative-sense, single-stranded RNA with 11,131 nucleotides. The genomic sequence contains 24.28%, 28.67%, 19.46% and 27.59% of G, A, T and C contents, respectively. It encodes 6 proteins as follows: a nucleoprotein (N), a phosphoprotein (P), a matrix protein (M), a glycoprotein (G), a non-virion protein (NV), and a polymerase (L)

(5) Genetic analysis of the 303 nucleotides encoding glycoprotein genes sub-categorise IHNV into five genotypes, namely U, L, E, J and M. The IHNV genotypes differ in geographic location, e.g. sockeye salmon in the Northeast Pacific – genotype U; chinook salmon in California, USA. – genotype L; and rainbow trout in Europe, Asia and Idaho, USA – genotypes E, J and M, respectively.

Clinical signs and pathological changes

(1) Lethargic or slow movement, while some have abnormal rapid swimming.

(2) Darkened skin, pale gills, exophthalmia, sticky excretions.

(3) Ascites and abdominal distention, petechiae on body.

(4) Fish may be anaemic with no intestinal contents upon post-mortem examination. Pale liver, kidney and spleen, with petechiae haemorrhage in internal organs.

(5) Haematopoietic tissues, spleen, kidney, liver, pancreas and gastrointestinal tract degeneration and necrosis, with necrosis of the eosinophils in the intestinal wall.

Diagnostic methods

(1) **Virus isolation** Use EPC and FHM cell lines and incubate at 15℃.

(2) **Reverse transcriptase-polymerase chain reaction (RT-PCR)** Use the primers IHNV-F (5′-AGA-GAT-CCC-TAC-ACC-AGA-GAC-3′) and IHNV-R (5′-GGT-GGT-GTT-GTT-TCC-GTG-CAA-3′) with annealing temperature at 50℃.The amplicon size is 693bp. Perform restriction enzyme digestion or sequencing on the amplicon for confirmation.

(3) **Enzyme-linked immunosorbent assay (ELISA)**

Preventive measures

(1) Maintain good hygiene management in aquaculture farms.

(2) Stamp out the pond to eradicate the pathogens.

(3) Supplement with oral immunostimulant and metabolic modifiers such as polysaccharides and multi-vitamins, etc.

(4) Use commercial vaccines.

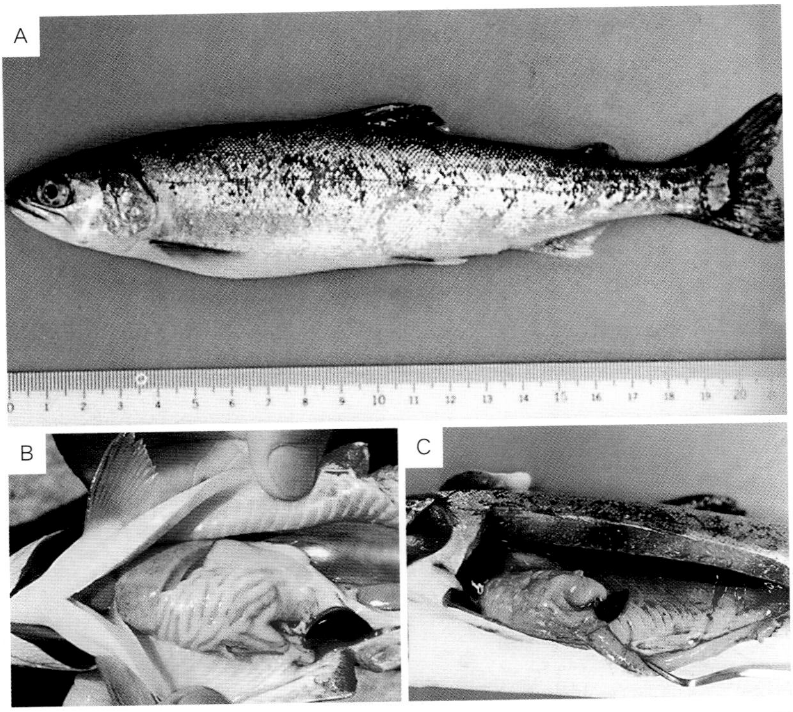

患传染性造血器官坏死的鱼

A．体色变黑，体表点状出血　B．解剖检查时可见到病鱼贫血的症状：鳃、肝脏褪色
C．在腹膜、脂肪组织及肝脏等处可见到点状出血和肌肉内出血等病变
[源自《新鱼病图鉴》，小川和夫]

Macroscopic findings of affected fish

A．Darkened body color with petechiae on the external body
B&C．Fish is anaemic with pale gill and liver. Petechial haemorrhages present in multiple organs upon post-mortem examination
[Source：*New Atlas of Fish Diseases* ，Kazuo Ogawa]

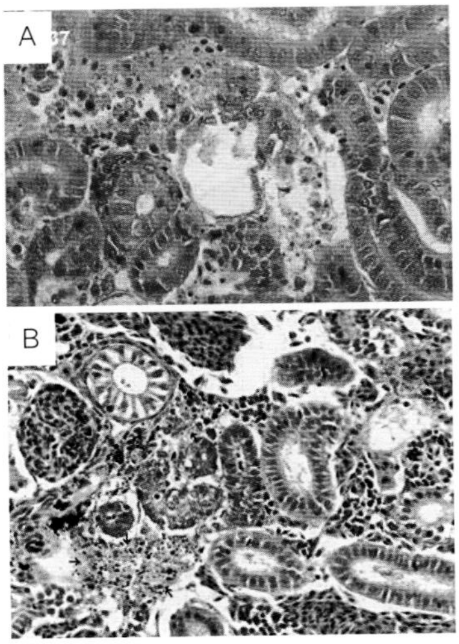

患病虹鳟细胞及组织病变（HE染色）

A．肾造血组织坏死
B．病鱼后肾，人工感染鲑粒弹状病毒后第三天造血组织（箭头所示）严重坏死
[源自 P. George和 Dr. Diane Elliott]

Histological lesions of affected fish(HE staining)

A．Severe necrosis present in haematopoietic organs
B．Metanephroi from affected fish, hematopoietic tissue (shown by arrow) is severely necrotic on the third day after challenge with *Salmonid novirhabdovirus*
[Source：P. George and Dr. Diane Elliott]

草鱼出血病

疾病概述

【概述】 草鱼出血病是淡水鱼类一种病毒性疾病，带有红鳍、红鳃、红肠和红肌肉等一种或多种症状。

【易感宿主】 草鱼（*Ctenopharyngodon idellus*）、青鱼（*Mylopharyngodon piceus*）、麦穗鱼（*Pseudorasbora parva*）、鲢（*Hypophthalmichthys molitrix*）、鳙（*Aristichthys nobilis*）、鲫（*Carassius auratus*）和鲤（*Cyprinus carpio*）。

【易感阶段】 体长2.5 ~ 15cm的草鱼和1龄的青鱼最易感，2龄以上草鱼和青鱼极少发病。

【发病水温】 水温20 ~ 33℃可发病，25 ~ 28℃是发病高峰期，死亡率达70%以上。

【地域分布】 流行于中国长江中下游以南地区以及越南。

病原

（1）病原为草鱼呼肠孤病毒（Grass carp reovirus，GCRV）。

（2）属呼肠孤病毒科（*Reoviridae*）、水生呼肠孤病毒属（*Aquareovirus*）、水生呼肠孤病毒C群（*Aquareovirus* C）。

（3）病毒粒子呈球状，直径70nm，有双层衣壳，无囊膜。

（4）含有11个双链RNA片段。不同毒株之间，病毒基因组的总分子质量差异不大，但是各片段分子质量有所差别。按照分子质量大小，将11个片段分为大片段（L1、L2、L3）、中等片段（M1、M2、M3）和小片段（S1、S2、S3、S4、S5）。

（5）根据已发表的毒株全基因组或部分节段基因序列，将目前检测到的GCRV毒株分为3个基因型（Ⅰ型、Ⅱ型和Ⅲ型）。

临床症状和病理学变化

（1）患病鱼在池塘边缘停滞，游动缓慢，体表发黑，眼突出。

（2）口腔、鳃盖、鳃和鳍条基部出血。

（3）剖检可见肌肉点状或块状出血、肠道出血，肝脏、脾脏和肾脏有不同程度的出血，肝脏失血发黄。

（4）根据临床症状不同，分为“红肌肉”“红肠”“红鳍红鳃”3种类型，患病鱼可出现一种或两种以上的症状。

（5）组织病理学观察，肝细胞退化、坏死，肝脏、脾脏内血管充血或出血。

诊断方法

（1）病毒分离 使用草鱼肾脏组织细胞系（CIK），分离温度为28℃。

（2）SDS-PAGE电泳 患病草鱼组织匀浆液或细胞培养病毒经离心纯化后，提取病毒基因组RNA，进行SDS-PAGE电泳。硝酸银染色观察，病毒基因组分节段RNA分子呈现11个条带，则可判定为感染草鱼呼肠孤病毒。

（3）RT-PCR 引物为GCRV-F（5′- TAY-GTV-ACM-SCC-MGR-GGW-GG-3′）和GCRV-R（5′-AAD-TGY-TGY-ACC-ATG-DYC-TGC-3′），退火温度为60℃，扩增产物长度为590bp（Ⅰ型）、590bp或593bp（Ⅱ型）、587bp（Ⅲ型），扩增产物经测序后进行判定。该方法适用于3种基因型草鱼呼肠孤病毒的检测。

防治方法

（1）水深1m时，每亩*池塘生石灰用量为125kg；排干池水（塘底留水10cm）时，生石灰用量为每亩60 ~ 80kg。清塘后7 ~ 10d、水温15 ~ 20℃时放养鱼种。

（2）晴天中午开动增氧机，减少底层氧债，改善池水溶解氧状况。

（3）对养殖场引入的亲鱼或苗种，采取严格的检疫和隔离措施。

（4）保持养殖场良好的卫生管理水平，定期向池塘施用微生态制剂，改良池塘底质和水质。水产上常用的微生态制剂主要有光合细菌、芽孢杆菌、硝化细菌、反硝化细菌、益生菌（EM菌）等，用来降低水体氨氮、亚硝酸盐，抑制病原菌繁殖等。

（5）清塘消毒，杀灭病原体。

（6）内服免疫增强与代谢调节剂，如多糖、多肽、多种维生素等。

（7）免疫预防，使用草鱼出血病灭活疫苗，通过浸泡、口服或注射途径进行免疫预防。鱼种放养前，对规格50 ~ 100g的草鱼种，人工注射草鱼出血病灭活疫苗。考虑到草鱼出血病常与草鱼“三病”（赤皮、肠炎、烂鳃病）并发的特点，可将草鱼出血病灭活疫苗和草鱼“三病”三联疫苗同时注射。为防止注射导致的细菌性感染，可在注射液中添加抗生素。

（8）药物预防，采用复方二硫氰甲烷1 ~ 1.5g/kg拌饵投喂，连用5 ~ 6d。发病草鱼池，用10%聚维酮碘（用量为45 ~ 75mg/m^3）或将20%戊二醛用水稀释300 ~ 500倍（用量为40mg/m^3）遍洒，隔日1次，连用2次。

（9）天然植物抗病毒药物治疗，如大黄、板蓝根、鱼腥草、黄芪等，超微粉碎后再拌食投喂，预防和治疗效果明显。

* 1亩≈667m^2，下同。——编者注

Hemorrhagic disease of grass carp

Disease overview

[Disease Characteristic] A viral disease affecting freshwater fish, with one or more of the clinical signs of red fins, red gills, red intestines and red muscles etc.

[Susceptible Host] Infect grass carp (*Ctenopharyngodon idellus*), black carp (*Mylopharyngodon piceus*), stone moroko (*Pseudorasbora parva*), silver carp (*Hypophthalmichthys molitrix*), bighead carp (*Aristichthys nobilis*), goldfish (*Carassius auratus*) and common carp (*Cyprinus carpio*).

[Susceptible Stage] Grass carp with a body length of 2.5~15cm and black carp of 1-year-old are the most susceptible stages. Grass carp and black carp more than two years of age seldom show any clinical disease.

[Outbreak Water Temperature] Disease outbreaks can occur when the water temperature is at 20~33℃ , with peak of outbreaks occurring at 25 ~ 28℃ . The mortality rate can be over 70%.

[Geographic Distribution] Endemic in areas of the south of the middle and lower reaches of the Yangtze River in China and Vietnam.

Aetiological agent

(1) Grass carp reovirus (GCRV).

(2) Family: *Reoviridae*. Genus: *Aquareovirus*. Group: *Aquareovirus* C.

(3) The virus is spherical in shape, 70nm in diameter, with double layers of capsids and no capsule.

(4) Viral genome contains 11 double-stranded RNA segments. The total molecular weights of the viral genomes have no significant difference among different viral strains, but there is difference among the molecular weights of each segment. According to molecular sizes, the 11 segments are classified into large segments (L1, L2, L3), intermediate segments (M1, M2, M3) and small segments (S1, S2, S3, S4, S5).

(5) According to the published viral complete genome or partial segmental gene sequences, the currently detected GCRV viral strains are classified into 3 genotypes (genotype Ⅰ, genotype Ⅱ and genotype Ⅲ).

Clinical signs and pathological changes

(1) Infected fish show stagnation and slow movement near the edge of the ponds, darkened body surface and exophthalmia.

(2) Haemorrhage of the oral cavity, gill operculum, gills, and fin bases.

(3) Post-mortem examination shows petechiae or ecchymosis in muscles, intestinal haemorrhage, and various degrees of haemorrhage in liver, spleen and kidneys. Blood loss in the liver with yellowish discoloration.

(4) Classified into 3 types of diseases according to the different clinical signs: "Red muscle" "Red intestine" and "Red fin red gills". Infected fish can show one or more of these clinical signs.

(5) Histologically, hepatocytes show degeneration and necrosis, congested or hemorrhagic blood vessels are present in liver and spleen.

Diagnostic methods

(1) **Virus isolation** Using grass carp kidney tissue cell line (CIK) and incubate at 28℃ .

(2) **SDS-PAGE electrophoresis** After centrifugation and purification of the tissue homogenate from infected grass carp or virus culture media, extract viral genomic RNA to perform SDS-PAGE electrophoresis. Stain with silver nitrate to observe. If the viral genomic segmental RNA molecule shows 11 bands, it can be identified as grass carp reovirus infection.

(3) **Reverse transcriptase-polymerase chain reaction (RT-PCR)** Use the primers GCRV-F (5′- TAY-GTV-ACM-SCC-MGR-GGW-GG-3′) and GCRV-R (5′-AAD-TGY-TGY-ACC-ATG-DYC-TGC-3′) with annealing temperature at 60℃. The amplicon size is 590bp (Type Ⅰ), 590bp or 593bp (Type Ⅱ), and 587bp (Type Ⅲ). Perform sequencing on the amplicons for confirmation. This method is applicable to test for all three genotypes of grass carp reoviruses.

Preventive measures

(1) For pond with water depth of 1m, use 125kg per 667m^2 quicklime for disinfection. After draining (leaving water depth of 10cm at the bottom of the pond), use 60~80kg per 667m^2 quicklime for disinfection. After stamping out the pond, stock fingerlings on the 7th~10th day with water temperature of 15~20℃ .

(2) At noon on sunny days, activate the aerator to reduce oxygen debt at the bottom of the ponds and to increase the dissolved oxygen level.

(3) Maintain strict quarantine and isolation measures for the broodstock or fry/fingerlings introduced to the aquaculture farm.

(4) Maintain good hygiene in the aquaculture farms. Regularly apply microecological agents

to the ponds to improve sediment and water quality. The microecological agents commonly used in aquaculture include photosynthetic bacteria, *Bacillus*, nitrifying bacteria, denitrifying bacteria, and effective microorganisms. These agents are used to reduce ammonia nitrogen and nitrite in water, and to inhibit the growth of pathogenic bacteria, etc.

(5) Stamp out the pond to eradicate the pathogen.

(6) Oral administration of immunostimulant and metabolic modifiers, such as polysaccharides, polypeptide, multivitamins, etc.

(7) Immunization. Use grass carp hemorrhagic disease vaccine for immunization and disease prevention through the route of immersion, oral administration or injection. Inject grass carp fingerlings of sizes 50~100g with inactivated grass carp hemorrhagic disease vaccine prior to stocking. Considering that grass carp hemorrhagic disease often occurs with the "three diseases" (Red skin, Enteritis, Gill rot disease) of grass carp concurrently, the grass carp hemorrhagic disease vaccine may also be injected simultaneously with the grass carp "three disease" trivalent vaccine. To prevent secondary bacterial infection due to injection, antibiotics can be added into the injection.

(8) Medication prophylaxis. Mix 1~1.5g/kg of compound methylene dithiocyanate in feed for 5~6 consecutive days. Splash infected pond with 10% povidone iodine at a dosage of 45~75mg/m^3, or 300~500 times water-diluted 20% glutaraldehyde at a dosage of 40mg/m^3 once every other day for two times.

(9) Natural herbal antiviral drug treatment. Mix natural herbal antiviral drugs, such as rhubarb, *Isatidis radix*, *Houttuynia cordata*, *Astragalus propinquus*, etc., with feed after grinding, which shows effective prevention and treatment.

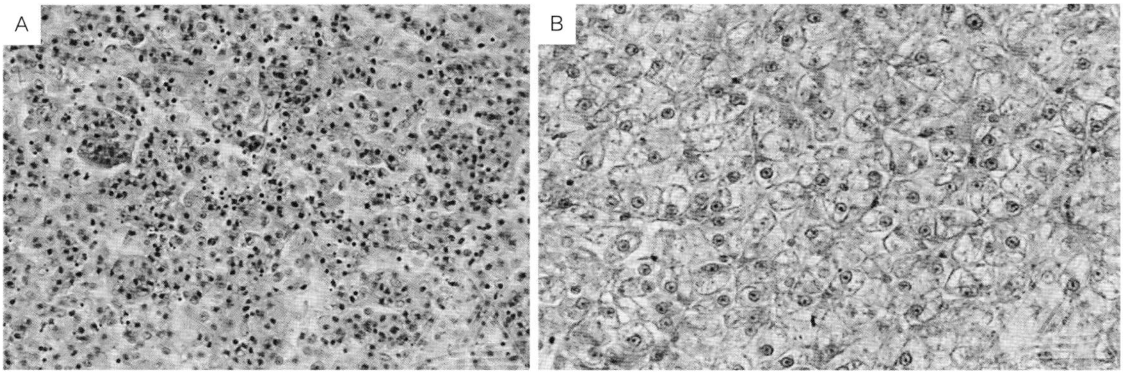

患草鱼出血病的鱼细胞及组织病变（HE染色）

A．脾脏几乎不见淋巴细胞，细胞散在坏死，严重瘀血　B．肝脏轻度水变性

[源自利洋*]

Histological lesions of affected fish (HE staining)

A．Spleen. Severe congestion, lymphoid depletion and sporadic necrosis　B．Liver. Hydropic degeneration of hepatocytes

[Source：LIYANG AQUATIC]

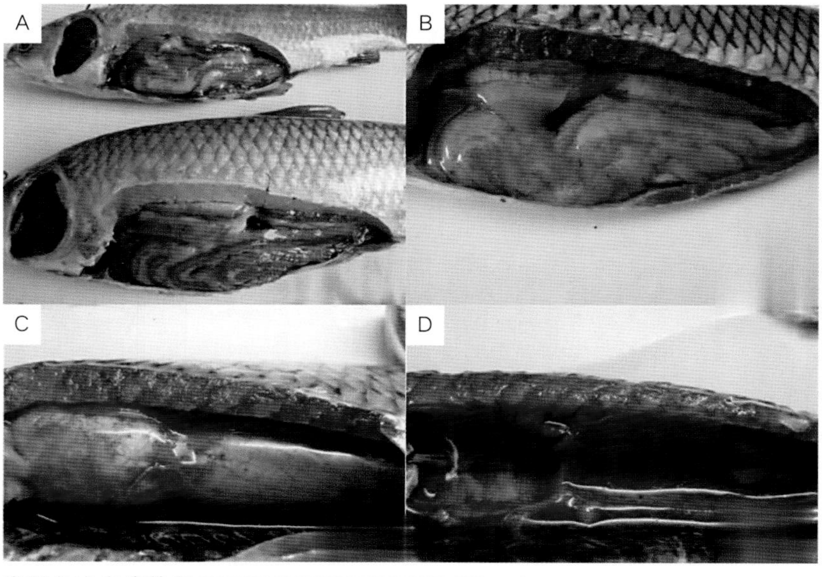

患草鱼出血病的鱼

A．烂鳃，肠系膜、肝脏、脾脏、鳔膜出血　B．肌肉出血，肝脏发黄，肠系膜脂肪出血

C．鱼鳔严重出血　D．肾脏肿大，有出血点

[源自唐绍林及利洋]

Macroscopic findings of affected fish

A．Gill necrosis. Haemorrhage present in the liver, spleen, swimming bladder and mesenteries

B．Pale liver. Haemorrhage of muscles and mesenteries　C．Severe haemorrhage of the swimming bladder　D．Kidney enlargement with petechiae

[Source：Shaolin Tang and LIYANG AQUATIC]

---

* 利洋：广州利洋水产科技股份有限公司。

病毒性出血性败血症

疾病概述

【概述】 病毒性出血性败血症是一种全身性出血败血的烈性传染病。

【易感宿主】 主要感染包括鲑科（Salmonidae）、狗鱼科（Esocidae）、鲱科（Clupeidae）、鳕科（Gadidae）、鲽科（Pleuronectidae）、菱鲆科（Scophthalmidae）、牙鲆科（Paralichthyidae）、水珍鱼科（Argentinidae）、胡瓜鱼科（Osneridae）、玉筋鱼科（Ammodytidae）、虾虎鱼科（Gobiidae）、海鲫科（Embiotocidae）、石首鱼科（Sciaenidae）、鲭科（Scombridae）、鲈科（Percidae）、狼鲈科（Moronidae）、刺鱼科（Gasterosteidae）、鲤科（Cyprinidae）和七鳃鳗科(Petromyzonidae）在内的80多种淡水和海水鱼类。宿主范围大，不同种类致病性存在较大差异。

【易感阶段】 可导致各年龄段的易感鱼发病和死亡。

【发病水温】 可全年流行。9 ～ 12℃死亡率最高。

【地域分布】 20世纪80年代之前，仅限于欧洲大陆的养殖虹鳟发病。随后，在太平洋、北美洲大西洋沿岸和五大湖、英国周边海域、波罗的海、斯卡格拉克海峡和卡特加特海峡、日本海域以及黑海地区养殖与野生鱼类中检测和分离到病毒性出血性败血症病原。

【疾病地位】 世界动物卫生组织（OIE）将其列入水生动物疫病名录。

病原

（1）2017年，国际病毒分类委员会将病原名称改为鱼粒弹状病毒（Piscine novirhabdovirus）。过去病原名称为病毒性出血性败血症病毒（Viral haemorrhagic septicaemia virus，VHSV）。

（2）属弹状病毒科（*Rhabdoviridae*）、粒外弹状病毒属（*Novirhabdovirus*）。

（3）病毒呈弹状，长度约为180nm，直径约70nm，有囊膜。

（4）病毒基因组为不分节段、负义、单股的RNA，包含有11 158个核苷酸；编码6个蛋白，分别是核蛋白（N）、磷蛋白（P）、基质蛋白（M）、糖蛋白（G）、非结构蛋白（NV）和RNA依赖的RNA聚合酶（L）。

（5）对N基因和G基因全长或者部分基因片段进行分析，分为4个主要基因型：Ⅰ型，分为Ia ～ Ie 5个亚型，包括欧洲淡水分离株、黑海地区分离株和一组来自波罗的海、卡特加特海峡、斯卡格拉克海峡、北海和英吉利海峡的海水分离株；Ⅱ型，波罗的海分离株；Ⅲ型，北大西洋海到挪威海岸、北海、斯卡格拉克海峡和卡特加特海峡分离株；Ⅳ型，北美洲、日本和韩国分离株。

临床症状和病理学变化

（1）游动异常，不活泼。

（2）体表发黑，眼球突出，鳃发白。

（3）有腹水腹部膨胀，鳍、鳃、眼睛和体表出血。

（4）剖检可见表皮、肌肉及内脏器官有点状出血，特别是背部肌肉出血明显。急性感染期，肾脏暗红，并出现严重坏死。脾脏肿胀。肝脏呈灰白色或有斑点。胃肠道特别是后肠呈灰白色且无食物。

诊断方法

（1）病毒分离 使用蓝鳃鱼细胞系（BF-2）、鲤上皮瘤细胞系（EPC）和肥头鲤细胞系（FHM），分离温度为15℃。

（2）RT-PCR 引物为VN-F（5′-ATG-GAA-GGA-GGA-ATT-CGT-GAA-GCG-3′）和VN-R（5′-GCG-GTG-AAG-TGC-TGC-AGT-TCC-C-3′），退火温度为55℃，扩增产物长度为505bp，扩增产物经酶切或测序后进行进一步诊断。

（3）酶联免疫吸附试验（ELISA）

（4）Real-time RT-PCR 引物和探针有2组可选用：

①上游引物VHSV-1F（5′-AAA-CTC-GCA-GGA-TGT-GTG-CGT-CC-3′），下游引物VHSV-1R（5′-TCT-GCG-ATC-TCA-GTC-AGG-ATG-AA-3′），探针VHSV-1MGB（5′-FAM-TAG-AGG-GCC-TTG-GTG-ATC-TTC-TG-BHQ1-3′）。

②上游引物VHSV-2F（5′-ATG-AGG-CAG-GTG-TCG-GAG-G-3′），下游引物VHSV-2R（5′-TGT-AGT-AGG-ACT-CTC-CCA-GCA-TCC-3′），探针VHSV-2MGB（5′ FAM-TAC-GCC-ATC-ATG-ATG-AGT-MGBNFQ-3′）。

防治方法

（1）保持养殖场良好的卫生管理水平。

（2）加强引进苗种和亲本的检疫。

（3）对养殖水源、设施彻底消毒，杀灭病原体。

（4）内服免疫增强与代谢调节剂，如多糖、多种维生素等。

（5）注射商品化疫苗。

Viral haemorrhagic septicaemia, VHS

Disease overview

[Disease Characteristic] VHS is a potent infectious disease causing systemic haemorrhage and septicaemia.

[Susceptible Host] Mainly affect fish species in the families of Salmonidae, Esocidae, Clupeidae, Gadidae, Pleuronectidae, Scophthalmidae, Paralichthyidae, Argentinidae, Osneridae, Ammodytidae, Gobiidae, Embiotocidae, Sciaenidae, Scombridae, Percidae, Moronidae, Gasterosteidae, Cyprinidae and Petromyzonidae, including more than 80 freshwater and marine fish species. Wide host range, the pathogenicity varies greatly in different species.

[Susceptible Stage] Cause clinical disease and death in susceptible fish of all ages.

[Outbreak Water Temperature] VHS is endemic throughout the year with highest mortality rate at water temperature of 9~12℃ .

[Geographical Distribution] Before the 1980s, the disease was limited to cultured rainbow trout in the European continents. Subsequently, the pathogen of VHS were detected and isolated from cultured and wild fish in the Pacific Ocean, the Atlantic coast and the five Great Lakes of North America, as well as the UK sea area, the Baltic Sea, the Skagerrak Strait, the Kattegat Strait, the Japan sea area and the Black Sea.

[Disease Status] OIE-listed Aquatic Animal Disease.

Aetiological agent

(1) Viral haemorrhagic septicaemia virus (VHSV). International Committee on Taxonomy of Viruses renamed the pathogen to Piscine novirhabdovirus in 2017.

(2) Family: *Rhabdoviridae*. Genus: *Novirhabdovirus*.

(3) The virion is bullet-shaped, enveloped, approximately 180nm in length and about 70nm in diameter.

(4) The viral genome consists of a non-segmental, negative-sense, single-stranded RNA, containing 11,158 nucleotides encoding 6 proteins, namely nuclear protein (N), phosphoprotein (P), matrix protein (M), glycoprotein (G), non-structural protein (NV) and polymerase (L).

(5) Analysis of the full-length or partial gene segments of N gene and G gene categorizes VHSV into 4 major genotypes: Type Ⅰ– further divided into 5 subtypes Ⅰa~Ⅰe, including European freshwater isolate, the Black Sea region isolate, and a group of marine isolates from the Baltic Sea, the Kattegat

Strait, the Skagerrak Strait, the North Sea and the English Strait; Type Ⅱ – Baltic isolate; Type Ⅲ – North Atlantic to Norwegian coast isolates, North Sea, Skagerrak and Kartega Strait isolates; Type Ⅳ – North American, Japanese and Korean isolates.

Clinical signs and pathological changes

(1) Abnormal movement and inactive.

(2) Darkened body color, exophthalmia, pale gills.

(3) Ascites, abdominal distention, haemorrhages of fins, gills, eyes and body surface.

(4) Post-mortem examination shows petechiae on body surface, in muscles and internal organs, in particular prominent in the back muscles. In acute stage, the kidneys appear dark red, with haemorrhage and severely necrotized. Spleen enlargement. Grayish-white liver, or with necrotic foci. Empty intestines appearing grayish-white, especially hindgut.

Diagnostic methods

(1) Virus isolation Use Bluegill Fish (BF-2), Epithelioma Papulosum Cyprini (EPC) and Fathead Minnow (FHM) cell lines and incubate at 15℃ .

(2) Reverse transcriptase-polymerase chain reaction (RT-PCR) Use the primers VN-F (5′-ATG-GAA-GGA-GGA-ATT-CGT-GAA-GCG-3′) and VN-R (5′-GCG-GTG-AAG-TGC-TGC-AGT-TCC-C-3′) with annealing temperature at 55℃. The amplicon size is 505bp. Perform sequencing or restriction enzyme digestion on the amplicon for confirmation.

(3) Enzyme-linked immunosorbent assay (ELISA)

(4) Real-time reverse transcriptase-polymerase chain reaction (Real-time RT-PCR) There are two sets of primers and probes available.

① Forward primer VHSV-1F (5′-AAA-CTC-GCA-GGA-TGT-GTG-CGT-CC-3′) and reverse primer VHSV-1R (5′-TCT-GCG-ATC-TCA-GTC-AGG-ATG- AA-3′) with probe VHSV-1MGB (5′-FAM-TAG-AGG-GCC-TTG-GTG-ATC-TTC-TG-BHQ1-3′).

② Forward primer VHSV-2F (5′-ATG-AGG-CAG-GTG-TCG-GAG-G-3′) and reverse primer VHSV-2R (5′-TGT-AGT-AGG-ACT-CTC-CCA-GCA-TCC-3′), with probe VHSV-2MGB (5′-FAM-TAC-GCC-ATC-ATG-ATG-AGT- MGBNFQ-3′).

Preventive measures

(1) Maintain good hygiene in aquaculture farms.

(2) Apply stringent quarantine measures on broodstock, fry and fingerling introduced to the farm.

(3) Thoroughly disinfect aquaculture water sources and facilities to eradicate the pathogens.

(4) Oral administration of immunostimulant and metabolic modifiers, such as

polysaccharides, multivitamins, etc.

（5）Use commercial vaccine.

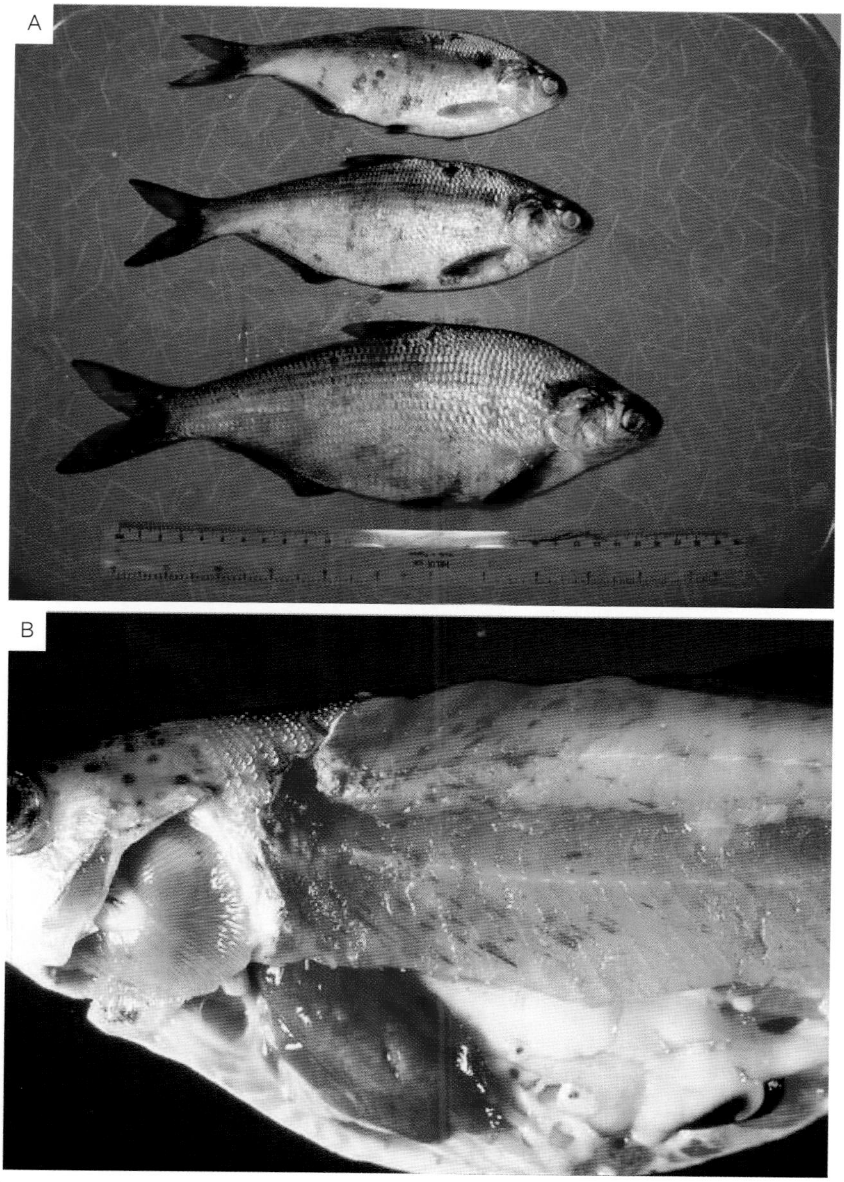

患VHS病鱼

A．体表有明显出血斑

B．肌肉及肠道明显出血，肝脏及鳃苍白

[源自Dr. P. Bowser]

Macroscopic findings of affected fish

A．Gizzard shad with external haemorrhage

B．Showing discoloration in the liver, haemorrhage in the muscles and intestines, and pale gill

[Source：Dr. P. Bowser]

流行性造血器官坏死病

疾病概述

【概述】 流行性造血器官坏死病是一种临床或者亚临床系统性感染的疾病。

【易感宿主】 自然情况下，仅感染河鲈（*Perca fluviatilis*）和虹鳟（*Oncorhynchus mykiss*）。浸泡感染证实下列鱼类会发生死亡：澳洲麦氏鲈（*Macquaria australasica*）、澳洲银鲈（*Bidyanus bidyanus*）、食蚊鱼（*Gambusia affinis*）和山南乳鱼（*Galaxias olidus*）。

【易感阶段】 虹鳟和河鲈在各个生长阶段均对该病毒易感。鱼苗和幼鱼比成鱼的症状明显。

【发病水温】 虹鳟为11 ～ 20℃。河鲈多在夏季发病。

【地域分布】 仅分布于澳大利亚。

【疾病地位】 世界动物卫生组织（OIE）将其列入水生动物疫病名录。

病原

（1）病原为流行性造血器官坏死病毒（Epizootic haematopoietic necrosis virus, EHNV）。

（2）属虹彩病毒科（*Iridoviridae*）、蛙病毒属（*Ranavirus*）。

（3）病毒呈二十面体结构，病毒粒子直径为150 ～ 180nm，无囊膜，病毒在细胞核和胞浆内复制，在细胞质中呈晶格状排列。

（4）病毒基因组为双链DNA，大小为150 ～ 170kb，主要衣壳蛋白（Major capsid protein，MCP）由分子质量48 ～ 55ku 的多肽链构成，占整个病毒粒子多肽的40% ～ 45%，*MCP* 基因在遗传学上高度保守。

（5）对病毒部分基因片段扩增产物进行测序分析或者进行酶切，可以区分与其同一个属的其他病毒，如玻勒虹彩病毒（Bohle virus, BIV）、欧鲖病毒（European catfish virus, ECV）、欧鲇病毒（European sheatfish virus, ESV）和桑蒂库帕蛙病毒（Santee-Cooper rana virus）。

临床症状和病理学变化

（1）患病鱼昏睡、顶水，鱼体失去平衡，鳃张开，体色发暗。

（2）体表无特异性损伤。

（3）肝脏、肾脏和脾脏肿大，肝脏上有白色或黄色坏死。

（4）肝脏、肾脏和脾脏可见急性的单个、多个或局部大片凝固性或液化性坏死病灶。靠近肝脏、肾脏坏死区域周围，有胞质内嗜碱性包涵体，在心脏、胰腺、胃肠道、鳃和假鳃中可见坏死性损伤。

诊断方法

(1)病毒分离 使用蓝鳃鱼细胞系(BF-2)、鲤上皮瘤细胞系(EPC)、肥头鲤细胞系(FHM)和草鱼性腺细胞系(GCO),22℃培养,如感染细胞出现致细胞病变效应(CPE),则可判定为疑似流行性造血器官坏死病感染。

(2)PCR 有3对引物可供选择:

① EHNV-M151 (5′-AAC-CCG-GCT-TTC-GGG-CAG-CA -3′) 和EHNV-M152 (5′-CGG-GGC-GGG-GTT-GAT-GAG-AT-3′),退火温度为50℃,扩增产物长度为321bp。

② EHNV-M153 (5′- ATG-ACC-GTC-GCC-CTC-ATC-AC-3′) 和EHNV-M154 (5′- CCA-TCG-AGC-CGT-TCA-TGA-TG -3′),退火温度为50℃,扩增产物长度为625bp。

③ EHNV-MCP-F (5′-CGC-AGT-CAA-GGC-CTT-GAT-GT-3′) 和EHNV-MCP-R (5′-AAA-GAC-CCG-TTT-TGC-AGC-AAA-C-3′),退火温度为55℃,扩增产物长度为580bp。

上述扩增产物都需经酶切或测序后判定。

(3)酶联免疫吸附试验(ELISA)

防治方法

(1)对养殖场引入的亲鱼或苗种,采取严格的检疫和隔离措施。

(2)保持养殖场良好的卫生管理水平。

(3)清塘消毒,杀灭病原体。

(4)加强饲养管理等综合措施。

(5)培育和引进抗病品种,提高抗病能力。

Epizootic haematopoietic necrosis, EHN

Disease overview

[Disease Characteristic] It is a clinical or subclinical systemic infectious disease.

[Susceptible Host] Under natural conditions, only infect European perch (*Perca fluviatilis*) and rainbow trout (*Oncorhynchus mykiss*). Experimental infection has shown to be possible and resulted in death in the following fish species: Macquarie perch (*Macquaria australasica*), silver perch (*Bidyanus bidyanus*), western mosquitofish (*Gambusia affinis*) and mountain galaxias (*Galaxias olidus*).

[Susceptible Stage] Rainbow trout and European perch are susceptible to the virus at all ages. Fry and young fish show more prominent clinical signs than adults.

[Outbreak Water Temperature] Disease outbreaks occur at 11~20℃ for rainbow trout. The outbreaks in European perch usually occur in summer.

[Geographic Distribution] Only distributed in Australia.

[Disease Status] OIE-listed Aquatic Animal Disease.

Aetiological agent

(1) Epizootic haematopoietic necrosis virus (EHNV).

(2) Family: *Iridoviridae*. Genus: *Ranavirus*.

(3) The virus possesses an icosahedral structure, 150~180nm in diameter, non-capsulated. It replicates inside the nucleus and cytoplasm, with a lattice arrangement in the cytoplasm.

(4) The viral genome is a double-stranded DNA with a size of 150~170kb. The major capsid protein (MCP) is formed by a polypeptide chain with a molecular weight of 48~55ku, which accounts for 40%~45% of the entire virion polypeptide. The *MCP* gene is highly conserved genetically.

(5) Sequencing or enzymatic cleavage of amplicons of partial viral gene fragments can distinguish EHNV from other similar viruses of the same genus, e.g. Bohle virus (BIV), European catfish virus (ECV), European sheatfish virus (ESV) and Santee-Cooper rana virus.

Clinical signs and pathological changes

(1) Lethargic, swim towards water surface, lost body balance, opened gills and darkened body color.

(2) No specific lesion on the body surface.

(3) Swollen liver, kidney and spleen. Whitish or yellowish necrosis in the liver.

(4) Acute, single, multiple or locally extensive coagulative or liquefying necrotic lesions in the liver, kidney and spleen. Intracytoplasmic basophilic inclusions close to the necrotic areas of liver and kidney. Necrotic lesions can be seen in the heart, pancreas, gastrointestinal tract, gills and pseudobranchia.

Diagnostic methods

(1) Virus isolation Use Bluegill Fish (BF-2), Epithelioma Papulosum Cyprini (EPC), Fathead Minnow (FHM) and Grass Carp Ovary (GCO) cell lines and incubate at 22℃. If the infected cells show typical cytopathic effects (CPE), it can be suspected as EHNV infection.

(2) Polymerase chain reaction (PCR) There are 3 pairs of primers available.

① EHNV-M151 (5′-AAC-CCG-GCT-TTC-GGG-CAG-CA-3′) and EHNV-M152 (5′-CGG-GGC-

GGG-GTT-GAT-GAG-AT-3′) with annealing temperature at 50℃. The amplicon size is 321bp.

② EHNV-M153(5′-ATG-ACC-GTC-GCC-CTC-ATC-AC-3′) and EHNV-M154 (5′- CCA-TCG-AGC-CGT-TCA-TGA-TG -3′) with annealing temperature at 50℃. The amplicon size is 625bp.

③ EHNV-MCP-F(5′-CGC-AGT-CAA-GGC-CTT-GAT-GT-3′) and EHNV-MCP-R (5′-AAA-GAC-CCG-TTT-TGC-AGC-AAA-C-3′) with annealing temperature at 50℃ . The amplicon size is 580bp.

Perform sequencing on the amplicons for confirmation.

(3) Enzyme-linked immunosorbent assay (ELISA)

Preventive measures

(1) Maintain strict quarantine and isolation measures for broodstock, fry or fingerling introduced to the aquaculture farms.

(2) Maintain good hygiene in the aquaculture farms.

(3) Stamp out the pond to eradicate the pathogen.

(4) Strengthen aquaculture management and other integrated measures.

(5) Develop and introduce disease-resistant species to enhance disease resistance.

受感染鱼的临床症状

A. 受感染的幼鱼腹胀，死亡率高

B. 鳃坏死

[源自 J. Humphrey]

Clinical signs of affected fish

A. Affected fish farm shows high mortality in juvenile fish with abdominal distension

B. Gill haemorrhage in affected fish

[Source: J. Humphrey]

锦鲤疱疹病毒病

疾病概述

【概述】 锦鲤疱疹病毒病是一种严重传染性、急性出血性疾病。

【易感宿主】 仅感染鲤（*Cyprinus carpio carpio*）、锦鲤（*Cyprinus carpio*）及鲤杂交品种。

【易感阶段】 所有年龄阶段均易感。

【发病水温】 发病温度16 ~ 25℃，感染鱼群发病率可到100%，死亡率可达70% ~ 80%。常见二次感染，或并发细菌或寄生虫感染。

【地域分布】 欧洲的德国、英国等多个国家，亚洲的中国、印度尼西亚、日本、韩国、马来西亚、新加坡、泰国以及以色列，南非、美国和加拿大均有报道。

【疾病地位】 世界动物卫生组织（OIE）将其列入水生动物疫病名录。

病原

（1）2008年，国际病毒分类委员会将病原名称改为鲤疱疹病毒3型（Cyprinid herpesvirus 3）。过去病原名称为锦鲤疱疹病毒（Koi herpesvirus, KHV）。

（2）属异疱疹病毒科（*Alloherpesviridae*）、鲤鱼病毒属（*Cyprinivirus*）。

（3）病毒呈二十面体结构、核衣壳直径100 ~ 110nm，病毒颗粒直径167 ~ 200nm，含脂质囊膜以及位于核衣壳与囊膜间的无定形蛋白质。

（4）病毒基因组为线性双股DNA，包含1个大的中央区和2个22kb、分布于两端的重复区。病毒基因组编码ORF2、肿瘤坏死因子受体（TNFR）、ORF22、ORF25和RING等5个基因型家族。ORF81编码3型膜蛋白，是KHV最具有免疫原性的膜蛋白之一。

（5）对病毒基因组进行分析，分为日本分支J系、美国/以色列分支U/I系2个分支。欧洲、印度尼西亚还存在另外2个分支。

临床症状和病理学变化

（1）发病鱼停止游动，嗜睡，离群或聚集在入水口或池塘边，有些身体失去平衡，无方向感或躁动不安。

（2）皮肤上出现白斑和水泡，鳃苍白、出血，组织坏死，鳞片有血丝，体表发白或发红，有些部位粗糙，局部或体表全部上皮脱落，黏液分泌过度或不足，眼凹陷，体表和鳍条基部出血或糜烂。

（3）鳃组织出现炎症和坏死，初级鳃小片受侵蚀、次级鳃小片融合，初级和次级鳃小片顶部肿胀，鳃上皮增生和肥厚。肾脏、脾脏、胰腺、肝脏、脑、肠和口腔上皮出现炎症、坏死和核内包涵体。

（4）内脏病变包括腹腔粘连，内脏的色泽改变（变深或变浅）。肾脏和肝脏膨大，出现点状出血。

诊断方法

（1）病毒分离 使用锦鲤鳍条组织细胞系（KF-1）或鲤脑细胞系（CCB），分离温度为20℃。

（2）酶联免疫吸附试验（ELISA）

（3）PCR 有2对引物可供选择：

① KHV-TK-F（5′- GGG-TTA-CCT-GTA-CGA-G-3′）和KHV-TK-R（5′-CAC-CCA-GTA-GAT-TAT-GC-3′），退火温度为52℃，扩增产物长度为409bp。

② KHV-Sph-F（5′- GAC-ACC-ACA-TCT-GCA-AGG-AG-3′）和KHV-Sph-R（5′- GAC-ACA-TGT-TAC-AAT-GGT-CGC-3′），退火温度为63℃，扩增产物长度为292bp。

扩增产物经测序后进行判定。

（4）Real-time PCR（定量PCR） 引物为KHV-86f（5′-GAC-GCC-GGA-GAC-CTT-GTG-3′）和KHV-163r（5′-CGG-GTT-CTT-ATT-TTT-GTC-CTT-GTT-3′），探针KHV-109p（5′-CTT-CCT-CTG-CTC-GGC-GAG-CAC-G-3′），退火温度为60℃，扩增产物长度为78bp。

防治方法

（1）对养殖场引入的对抗特定病原（SPR）和无特定病原（SPF）亲鱼或苗种，采取严格的检疫和隔离措施。

（2）保持养殖场良好的卫生管理水平。

（3）清塘消毒，杀灭病原体。

（4）加强饲养管理等综合措施。

（5）培育和引进抗病品种，提高抗病能力。

Koi herpesvirus disease, KHVD

Disease overview

[Disease Characteristic] A severe contagious and acute haemorrhagic disease.

[Susceptible Host] Only affects common carp (*Cyprinus carpio carpio*) and Koi (*Cyprinus carpio*), and hybrid species of carps.

[Susceptible Stage] All age groups are susceptible.

[Outbreak Water Temperature] Disease outbreaks occur at 16~25℃. Mortality and morbidity rates can reach 70%~80% and 100%, respectively, in infected fish populations. Secondary or concurrent infection by bacterial and parasitic are common.

[Geographic Distribution] Reported in many countries in Europe such as Germany and England, in Asia such as China, Indonesia, Japan, Korea, Malaysia, Singapore, Thailand, and Israel, as well as in South Africa, America and Canada.

[Disease Status] OIE-listed Aquatic Animal Disease.

Aetiological agent

(1) Koi herpesvirus (KHV), renamed as Cyprinid herpesvirus by the International Committee on Taxonomy of Viruses (ICTV) in 2008.

(2) Family: *Alloherpesviridae*. Genus: *Cyprinivirus*.

(3) The virus has an icosahedral structure of 167~200nm in diameter, the nucleocapsid is 100~110nm in diameter. The viral particle has a lipid envelope coated with viral protein with amorphous protein between nucleocapsid and the envelope.

(4) The viral genome consists of a linear double stranded DNA with a large central unit and two duplicate zones of 22kb at each end. It encodes for five genotypes: ORF2, TNFR(tumor necrosis factor receptor), ORF22, ORF25 and RING. ORF81 encodes for type 3 membrane protein, one of the strongest immunogenicity of KHV.

(5) Genomic analysis subdivides KHV into two lineages: the Japanese J lineage and the America/Israel U/I lineages. There are two other lineages, namely the European and Indonesia lineages.

Clinical signs and pathological changes

(1) Clinically infected fish are inactive and lethargic, lost balance and disorientated or restless. They may be segregated, or gathered at the side of the pond or water inlet.

(2) White spots or blisters on the skin. Pale gill with haemorrhage and necrosis. Pale or reddened body surface with blood strands on scales. Some body parts become rough, with partial or complete detachment of epidermis. Over or under-secretion of mucus may be present. Enophthalmos, and haemorrhages on the body surface and bases of the fins, sometimes with erosion.

(3) Inflammation and necrosis of the gill. Erosion of primary lamellae, fusion of secondary lamellae, and swelling at the tips of the primary and secondary lamella. Hyperplasia and hypertrophy of gill epithelium. Inflammation, necrosis and intranuclear inclusions present in kidney,

spleen, pancreas, liver, brain, intestines and oral epithelium.

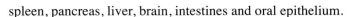

(4) Abdominal adhesions with or without discoloration of internal organs (paling or darkening). Enlarged kidney or liver with petechial haemorrhages.

Diagnostic methods

(1) **Virus isolation** Use Koi-fin -1 (KF-1) cell line or Common Carp Brain (CCB) cell line and incubate at 20℃ .

(2) **Enzyme-linked immunosorbent assay (ELISA)**

(3) **Polymerase chain reaction (PCR)** There are two pairs of primers available.

① KHV-TK-F (5′- GGG-TTA-CCT-GTA-CGA-G-3′) and KHV-TK-R (5′-CAC-CCA-GTA-GAT-TAT-GC-3′) with annealing temperature at 52℃ . The amplicon size is 409bp.

② KHV-Sph-F (5′- GAC-ACC-ACA-TCT-GCA-AGG-AG-3′) and KHV-Sph-R (5′- GAC-ACA-TGT-TAC-AAT-GGT-CGC-3′) with annealing temperature at 63℃. The amplicon size is 292bp.

Perform sequencing of the amplicons for confirmation.

(4) **Real-time polymerase chain reaction (qPCR)** Use the forward primer KHV-86f (5′-GAC-GCC-GGA-GAC-CTT-GTG-3′) and the reverse primer KHC-163r (5′-CGG-GTT-CTT-ATT-TTT-GTC-CTT-GTT-3′) with probe KHV-109p (5′-CTT-CCT-CTG-CTC-GGC-GAG-CAC-G-3′). The annealing temperature is 60℃ with amplicon size of 78bp.

Preventative measures

(1) Introduce Specific Pathogen Resistant (SPR) or Specific Pathogen Free (SPF) broodstock, fry or fingerling and apply stringent quarantine and isolation measures.

(2) Maintain good hygiene in the aquaculture farms.

(3) Stamp out the pond to eradicate the pathogens.

(4) Strengthen aquaculture management and other integrated measures.

(5) Develop and introduce disease-resistant species to enhance disease resistibility.

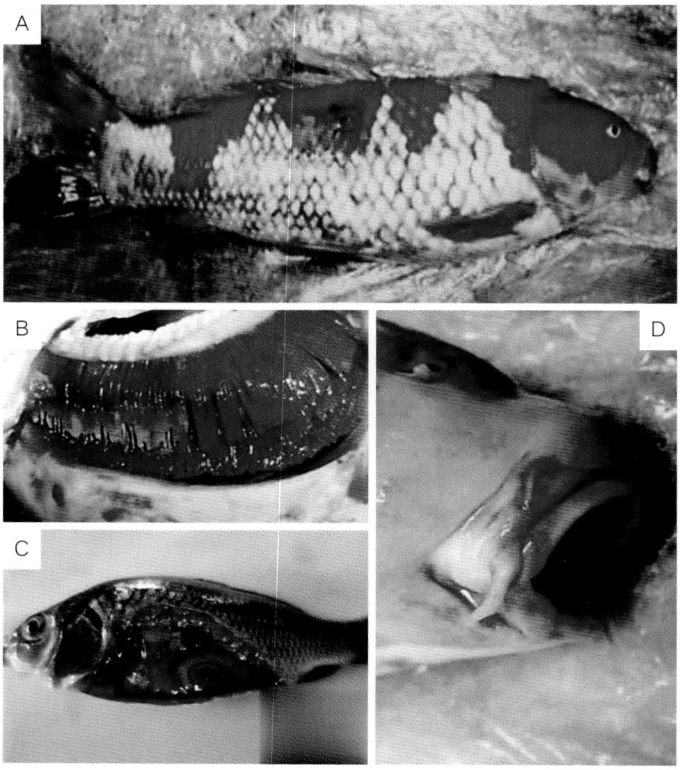

患病锦鲤症状

A. 严重皮肤溃疡，全身多处充血、出血

B. 鳃丝腐烂、出血

C. 发病鱼眼睛凹陷、烂鳃、肠道发红

D. 嘴明显充血、出血

[源自利洋及江育林]

Macroscopic findings of affected fish

A. Severe skin ulcerations with scale loss. Hyperaemia and haemorrhage of the body trunk and tail

B. Gill filament necrosis and haemorrhage

C. Enophthalmos, gill necrosis and intestinal haemorrhages

D. Marked hyperaemia and haemorrhage of the mouth

[Source: LIYANG AQUATIC and Yulin Jiang]

金鱼造血器官坏死病

疾病概述

【概述】 金鱼造血器官坏死病也称为疱疹病毒性造血器官坏死病（Herpesviral haematopoietic necrosis，HVHN），是鲫的一种高致病性病毒病。

【易感宿主】 仅感染金鱼（*Carassius auratus*）、鲫（*Carassius auratus*）及其普通变种，金鱼和鲤的杂交体也能感染CyHV-2而成为该病毒的携带者。

【易感阶段】 鱼卵、鱼苗、鱼种和亲鱼均可感染，但幼鱼更易感，死亡率可达100%。小于1龄的幼鱼有暴发性死亡现象。

【发病水温】 感染主要发生于春秋季节，受水温影响，15 ～ 25℃易发病。水温高于25℃时，发病率降低。

【地域分布】 该病在中国、澳大利亚、新西兰、日本、英国、美国等国家均有分布。

病原

（1）病原原名为金鱼造血器官坏死病毒（Goldfish haematopoietic necrosis virus，GFHNV），国际病毒系统分类与命名委员会(ICTV)将其命名为鲤疱疹病毒2型（Cyprinid herpesvirus 2，CyHV-2）。

（2）属异疱疹病毒科(*Alloherpesviridae*)、鲤鱼病毒属(*Cyprinivirus*)。

（3）一种双链DNA病毒，直径为100 ～ 110nm。外形呈二十面体，有囊膜的病毒呈椭圆形，直径为175 ～ 200nm，核衣壳呈六角形或球形。

（4）其基因组大小约290.3kb，与另外2种分离自鲤科鱼类的疱疹病毒，鲤痘病毒1型（Cyprinid herpesvirus 1，CyHV-1）和锦鲤疱疹病毒3型（Cyprinid herpesvirus 3，CyHV-3）均属于鲤疱疹病毒属，与斑点叉尾鮰病毒1型（Ictalurid herpesvirus 1，IcHV-1）关系相对较远。

临床症状和病理学变化

（1）病鱼精神沉郁，昏睡，食欲不佳或厌食，呼吸频率增加，停留在池塘或水箱底部。

（2）病鱼鳃出血后发白，鳔出血斑，鳍上有水泡状脓疱。有些病鱼腹部膨大，眼球突出。

（3）解剖后可见脾和肾肿胀呈苍白色，偶尔能见多处白色病灶，肝苍白，肠道空。

（4）组织病理表现为鳃小片融合，心脏病灶性坏死，口、表皮、肾、脾、胰腺、肠道由多病灶发展到弥漫性坏死，上皮细胞增生，头肾和体肾中造血细胞明显核固缩和核裂解。

诊断方法

（1）病毒分离 使用鲤上皮瘤细胞系（EPC）、胖头鲅细胞系（FHM）和罗非鱼卵巢细胞系（TO-2），分离温度为20℃。但该病毒对以上细胞系敏感性较低，且连续传代病毒滴度逐渐降低。

（2）PCR 引物有2组可供选择：

① CyHVpol-F（5′-CCC-AGC-AAC-ATG-TGC-GAC-GG-3′）和CyHVpol-R（5′-CCG-TAR-TGA-GAG-TTG-GCG-CA-3′），退火温度为55℃，扩增产物大小362bp。

② GFHNV F1（5′- GGA-CTT-GCG-AAG-AGT-TTG-ATT-TCT-AC-3′）和GFHNV R1（5′-CCA-TAG-TCA-CCA-TCG-TCT-CAT-C-3′），退火温度为60℃，扩增产物长度为366bp。

以上扩增产物经测序后进行判定。

（3）Real-time PCR 引物为GFHNV rtF1（5′-TCG-GTT-GGA-CTC-GGT-TTG-TG-3′）和GFHNV rtR1（5′-CTC-GGT-CTT-GAT-GCG-TTT-CTT-G-3′），探针为GFHNV rt（5′-FAM-CCG-CTT-CCA-GTC-TGG-GCC-ACT-ACC-BHQ1-3′），退火温度为59℃，Ct值为24。

防治方法

（1）对养殖场引入的亲鱼或苗种，采取严格的检疫和隔离措施。

（2）保持养殖场良好的卫生管理水平。

（3）选用SPR和SPF的苗种。

（4）切忌使用强氯精等刺激性强、杀藻力强的缓释氯制剂。杀灭藻类较重的杀虫药物也会刺激暴发该病。

（5）内服免疫增强与代谢调节剂，如多糖、多肽、多种维生素等。

Goldfish haematopoietic necrosis, GFHN

Disease overview

[Disease Characteristic] Highly infectious viral disease of *Carassius auratus*, also named Herpesviral haematopoietic necrosis, HVHN.

[Susceptible Host]Only affect goldfish (*Carassius auratus*) and crucian carp (*Carassius auratus*), as well as their common variants. Hybrids of goldfish and common carp can also be infected and become carriers.

[Susceptible Stage] Fish egg, fry, fingerling and broodstock are all susceptible to infection, but juveniles are more susceptible than adult fish. Juveniles less than one year old are prone to have outbreak with a mortality rate reaching 100%.

[Outbreak Water Temperature] Disease outbreaks mostly occur in spring and autumn, and are dependent on water temperature. Fish is prone to clinical infection at 15~25℃ , morbidity reduces when water temperature is higher than 25℃ .

[Geographic Distribution] The disease is widely distributed globally such as in China, Australia, New Zealand, Japan, England and America, etc.

Aetiological agent

(1) Goldfish haematopoietic necrosis virus, GFHNV. The International Committee on Taxonomy of Viruses (ICTV) renamed it as Cyprinid herpesvirus 2, CyHV-2.

(2) Family: *Alloherpesviridae*. Genus: *Cyprinivirus*.

(3) Double-stranded DNA virus, icosahedral in shape of 100~110nm in diameter. Enveloped virus appears ovoid in shape of 175~200nm in diameter, with a hexagonal or spherical nucleocapsid.

(4) The genome is approximately 290.3kb in size, and is very similar to the other two viruses in the same family, namely carp pox herpesvirus (Cyprinid herpesvirus 1, CyHV-1) and Koi herpesvirus (Cyprinid herpesvirus 3, CyHV-3), but less similar to channel catfish virus (Ictalurid herpesvirus 1, IcHV-1).

Clinical Signs and pathological changes

(1) Clinically affected fish will sink to the bottom of the pond and appear lethargic, with poor appetite or anorexic, as well as increased respiratory rate.

(2) Pale gills following haemorrhage of the gills, ecchymosis of the swim bladder, and blister and pustule on the fins of sick fish. Some fish may demonstrate exophthalmia and abdominal distension.

(3) Spleen and kidney are enlarged and pale in post-mortem examination. Occasionally, there are multiple whitish lesions, pale liver and empty intestines.

(4) Histologically, there are fusions of the lamellae in the gill, focal necrosis in the heart, multifocal lesions to diffuse necrosis in mouth, skin, kidney, spleen, pancreas and intestines. Hyperplasia of epithelium with significant nuclear pyknosis and karyorrhexis in the cells of head and body kidneys.

Diagnostic methods

(1) Virus isolation Use epithelioma papillosum cyprinid (EPC), fathead minnow cells (FHM) and tilapia ovary (TO-2) cell lines and incubate at 20℃. However, sensitivity is low with the above cell lines, with decreasing viral titres in successive passages.

(2) **Polymerase chain reaction** There are two pairs of primers.

① CyHVpol-F (5′-CCC-AGC-AAC-ATG-TGC-GAC-GG-3′) and CyHVpol-R(5′-CCG-TAR-TGA-GAG-TTG-GCG-CA-3′) with annealing temperature at 55℃. The amplicon size is 362bp.

② GFHNV F1 (5′-GGA-CTT-GCG-AAG-AGT-TTG-ATT-TCT-AC-3′) and GFHNV R1 (5′-CCA-TAG-TCA-CCA-TCG-TCT-CAT-C-3′) with annealing temperature at 60 ℃. The amplicon size is 366bp.

Perform sequencing on the amplicons for confirmation.

(3) **Real-time polymerase chain reaction** Use the primers GFHNV rtF1 (5′-TCG-GTT-GGA-CTC-GGT-TTG-TG-3′) and GFHNV rtR1 (5′-CTC-GGT-CTT-GAT-GCG-TTT-CTT-G-3′) with GFHNV rt probe (5′-FAM-CCGCTTCCAGTCTGGGCCACTACC-BHQ1-3′). The annealing temperature is 59℃, and the Ct value is 24.

Preventative measures

(1) Apply stringent quarantine measures on broodstock, fry and fingerling introduced to the farm.

(2) Maintain good hygiene management in aquaculture farms.

(3) Use SPR and SPF fry or fingerling for farming.

(4) Avoid using slow releasing chlorine agents with strong irritation or algaecidal effects. Pesticide with strong algaecide effect will also provoke outbreak of disease.

(5) Oral administration of immunostimulant and metabolic modifiers such as polysaccharides, peptides and multivitamins.

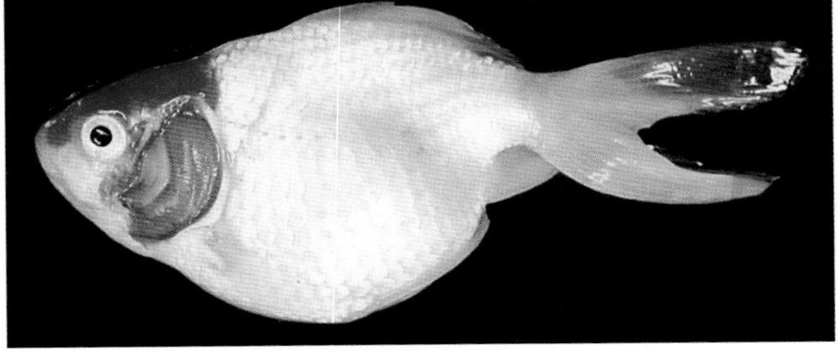

病鱼严重贫血使鳃明显褪色，外观无明显的特征性症状

[源自《新鱼病图鉴》，小川和夫]

Affected fish showing severe anaemia with pale gill without other apparent external lesion

[Source: *New Atlas of Fish Diseases*, Kazuo Ogawa]

组织学检查可见肾细胞坏死（HE染色）

[源自《新鱼病图鉴》，小川和夫]

Histological lesions of the kidney showing marked necrosis of the renal tubular structure (HE staining)

[Source: *New Atlas of Fish Diseases*, Kazuo Ogawa]

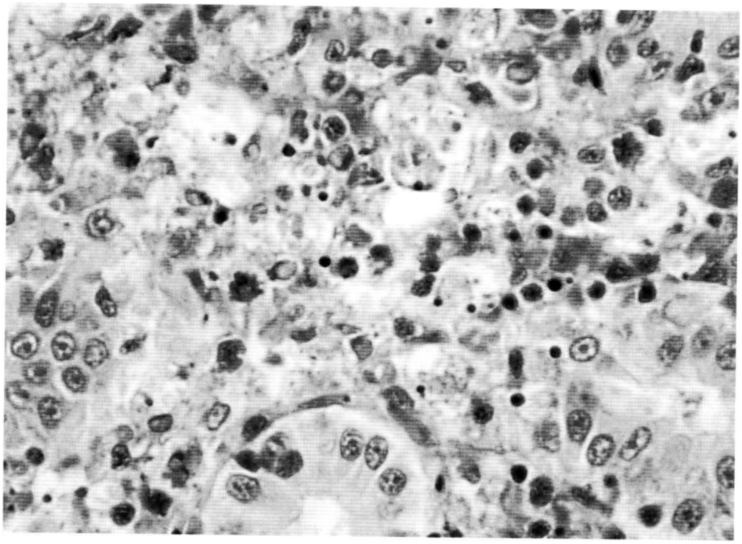

鲑传染性贫血病

疾病概述

【概述】 鲑传染性贫血病是一种大西洋鲑全身性致死性传染病。

【易感宿主】 自然条件下，感染大西洋鲑(*Salmo salar*)、虹鳟(*Oncorhynchus mykiss*)、银鲑(*Oncorhynchus* kisutch)、褐鳟(*Salmo trutta*)和鳟(*Squaliobarbus ourriculus*)。实验条件下，褐鳟、鳟、虹鳟、远东红点鲑(*Salvelinus leucomaenis*)、鲱(*Clupea pallasi*)和大西洋鳕(*Gadus morhua*)对其易感。

【发病水温】 一般情况下，全年各个时期都可能暴发鲑传染性贫血病（ISA）。夏初和冬季死亡率最高。在养殖场，ISA可能通过连续感染不同新个体，并不引起临床症状，从而导致一个群体的持续感染。

【地域分布】 在挪威、加拿大、英属法罗群岛、美国和智利的大西洋鲑中均有分布。爱尔兰的虹鳟和智利的银鲑也发生过ISA。

【疾病地位】 世界动物卫生组织（OIE）将其列入水生动物疫病名录。

病原

（1）病原为传染性鲑贫血症病毒（Infectious salmon anaemia virus, ISAV）。

（2）属正黏病毒科（*Orthomyxoviridae*）、传染性鲑贫血症病毒属（*Isavirus*）。

（3）病毒为单链、负股RNA，呈球形，直径100 ~ 300 nm，有囊膜。

（4）由8条负链RNA片段组成，其基因编码至少10种蛋白，现已鉴定4种主要结构蛋白，分别为68ku的核蛋白、22ku的基质蛋白、42ku的具有受体结合及破坏活性的血凝素酯酶蛋白（HE）和56ku的具有融合活性的表面糖蛋白，分别由第3、第8、第6和第5条基因片段编码。基因片段1、2和4分别编码病毒聚合酶PB1、PB2和PA。

（5）根据HE基因5′端差异，ISAV被分为欧洲群和北美群2个主要群。欧洲群进一步分为3个亚群。

临床症状和病理学变化

（1）HPR缺失型ISAV对宿主具有致病性，HPR0 ISAV对宿主没有致病性，多为带毒感染。

（2）病鱼昏睡，贴近网箱壁。

（3）鳃部苍白（鳃部有瘀血除外），眼球突出。腹部膨胀，眼前房出血，皮肤出血（特别是腹部皮肤），鳞片囊水肿；中心静脉窦和鳃小片毛细血管充血，在鳃部形成红细胞血栓。

(4) 解剖可见病鱼消化道无食物，腹腔和心包腔有微黄色和浅血色液体，鱼鳔肿胀，内脏和腹膜壁层小范围出血，肝脏局灶性或弥漫性暗红色，表面可能覆盖薄的纤维层，肾脏边缘变圆、肿胀、呈暗红色，切口有血和液体流出。盲囊、中肠和后肠肠壁黏膜暗红，骨骼肌上有出血点。

诊断方法

（1）病毒分离 只能对HPR缺失型ISAV进行分离，HPR0 ISAV暂时无法进行病毒分离。使用SHK-1或ASK细胞系，分离温度为15℃。

（2）RT-PCR HPR0 ISAV可用RT-PCR进行检测，上游引物为（5′-GAC-CAG-ACA-AGC-TTA-GGT-AAC-ACA-GA-3′），下游引物为（5′-GAT-GGT-GGA-ATT-CTA-CCT-CTA-GAC-TTG-TA-3′），退火温度为56℃，扩增产物长度为304bp，扩增产物经测序后进行判定。

（3）Real-time PCR 可用来检测HPR缺失型ISAV和HPR0 ISAV，有2对引物和探针可供选择：

①上游引物为（5′-CAG-GGT-TGT-ATC-CAT-GGT-TGA-AAT-G-3′），下游引物为（5′-GTC-CAG-CCC-TAA-GCT-CAA-CTC-3′），探针为（5′-6FAM-CTC-TCT-CAT-TGT-GAT-CCC-MGBNFQ-3′），退火温度为60℃，扩增产物长度为155bp。

②上游引物为（5′-CTA-CAC-AGC-AGG-ATG-CAG-ATG-T-3′），下游引物为（5′-CAG-GAT-GCC-GGA-AGT-CGA-T-3′），探针为（5′-6FAM-CAT-CGT-CGC-TGC-AGT-TC-MGBNFQ-3′），退火温度为60℃，扩增产物长度为104bp。

（4）酶联免疫吸附试验（ELISA）

防治方法

(1) 保持养殖场良好的卫生管理水平。

(2) 对养殖场引入的亲鱼或苗种，采取严格的检疫和隔离措施。

(3) 选育不带病原的健康苗种。

(4) 内服免疫增强与代谢调节剂，如多糖、多种维生素。

(5) 使用商品化疫苗。

Infectious salmon anaemia, ISA

Disease overview

[Disease Characteristic] Systemic and fatal infectious disease in Atlantic salmon.

[Susceptible Host] Under natural condition, affect Atlantic salmon *(Salmo salar)*, rainbow trout (*Oncorhynchus mykiss)*, silver salmon (*O. kisutch)* , brown trout (*S. trutta)* and trout *(Squaliobarbus ourriculus)*. In experimental setting, brown trout, trout, rainbow trout, whitespotted char (*Salvelinus leucomaenis),* Pacific herring *(Clupea pallasii)* and Atlantic cod *(Gadus morhua)* are susceptible.

[Outbreak Water Temperature] Generally, disase outbreaks of ISA can occur all year round. The mortality rate is the highest in early summer and winter. In aquaculture farms, ISA can repeatedly infect new individuals without causing clinical signs, which can result in persistent infection within a herd.

[Geographical Distribution] Disease is reported in Atlantic salmon in Norway, Canada, United Kingdom, Faroe Islands of North Ireland, USA and Chile. It has also been found in the rainbow trout in Ireland and silver salmon in Chile.

[Disease Status] OIE-listed Aquatic Animal Disease.

Aetiological agent

(1) Infectious salmon anaemia virus (ISAV).

(2) Family: *Orthomyxoviridae*. Genus: *Isavirus*.

(3) Single-stranded, negative sense RNA, spherical in shape, enveloped, 100~300nm in diameter.

(4) The viral genome consists of 8 strands of negative-sense RNA fragments encoding at least 10 proteins, of which 4 of them are the major structural proteins, including a 68ku nucleoprotein, a 22ku matrix protein, a 42ku haemagglutinin-esterase (HE) protein responsible for receptor-binding and receptor-destroying activity, and a 56ku surface glycoprotein with putative fusion (F) activity, encoded by genome segments 3, 8, 6 and 5, respectively. Segments 1, 2, and 4 encode the viral polymerases PB2, PB1 and PA, respectively.

(5) ISAV is subcategorized into European and North American clades according to the difference at the 5′ end of the HE gene. The European clade is further divided into three sub-clades.

Clinical signs and pathological changes

(1) HPR-deleted ISV is pathogenic to its host. HPR0 ISAV is not pathogenic, but infected fish will become carriers.

(2) Lethargic and to stay near the wall of the net.

(3) Pale gills (except congestion in the gill), exophthalmia, abdominal distension, haemorrhage in the anterior eye chamber and skin haemorrhage (especially on the abdomen). Scale sac oedema. Congestion of central venous sinusoids and capillaries of small lamellae, with thrombosis in the gills.

(4) Empty gastrointestinal tract. Yellow or blood-tinged fluid in peritoneal and pericardial cavities. Oedema of the swim bladder. Small areas of haemorrhage of the visceral and parietal peritoneum. Focal or diffuse dark reddening of liver with thin fibrin layer on the surface. Kidney swollen and dark red in color with blood and liquid leaking out from the cut surface. The intestinal mucosa is dark reddened in the caecum, midgut and hindgut. Petechiae of skeletal muscles.

Diagnostic methods

(1) **Virus isolation** Currently only applicable for HPR-deleted ISAV, but not HPR0 ISAV. Use SHK-1 or ASK cell line and incubate at 15℃.

(2) **Reversed transcriptase-polymerase chain reaction (RT-PCR)** HPRo ISAV can be detected by RT-PCR. Use forward primer (5′-GAC-CAG-ACA-AGC-TTA-GGT-AAC-ACA-GA-3′) and reverse primer (5′-GAT-GGT-GGA-ATT-CTA-CCT-CTA-GAC-TTG-TA-3′) with annealing temperature at 56℃. The amplicon size is 304bp. Perform sequencing on the amplicon for confirmation.

(3) **Real-time polymerase chain reaction (real time PCR)** Used for detection of HPR-deleted ISAV and HPRo ISAV. There are 2 pairs of primers and probes available：

① Forward primer (5′-CAG-GGT-TGT-ATC-CAT-GGT-TGA-AAT-G-3′) and reverse primer (5′-GTC-CAG-CCC-TAA-GCT-CAA-CTC-3′), with probe (5′-6FAM-CTC-TCT-CAT-TGT-GAT-CCC-MGBNFQ-3′). The annealing temperature is 60℃ and the amplicon size is 155bp.

② Forward primer (5′-CTA-CAC-AGC-AGG-ATG-CAG-ATG-T-3′) and reverse primer (5′-CAG-GAT-GCC-GGA-AGT-CGA-T-3′) with probe (5′-6FAM-CAT-CGT-CGC-TGC-AGT-TC-MGBNFQ-3′). The annealing temperature is 60℃ and the amplicon size is 104bp.

(4) **Enzyme-linked immunosorbent assay (ELISA)**

Preventative measures

(1) Maintain good hygiene in aquaculture farms.

(2) Apply stringent quarantine measures on broodstock, fry and fingerling introduced to the farm.

(3) Select SPF fry or fingerling for farming.

(4) Oral administration of immunostimulant and metabolic modifiers, such as polysaccharides, polypeptide, multivitamins, etc.

(5) Use commercial vaccines.

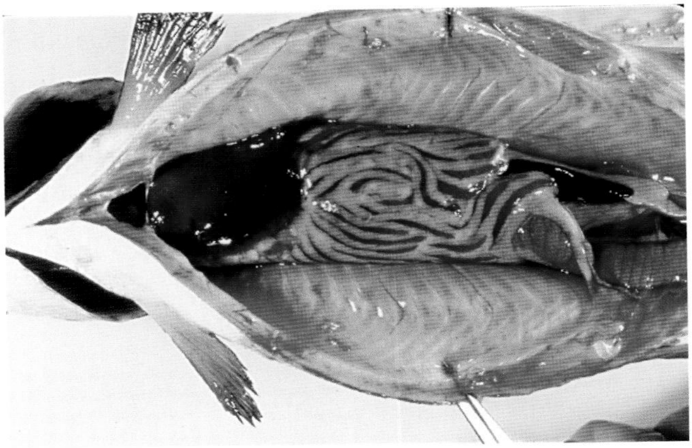

感染ISA的大西洋鲑可见肝脏颜色变暗，有腹水，脾肿大
[源自T Poppe]

Infected *Salmo salar*. Darkened liver, ascites and spleen enlargement
[Source：T Poppe]

鲑甲病毒病

疾病概述

【概述】 鲑甲病毒病又被称为鲑胰腺病（Pancreas disease，PD）或昏睡病（Sleeping disease，SD）。

【易感宿主】 不同基因型病毒的宿主有所不同，主要感染大西洋鲑（*Salmo salar*）、虹鳟（*Oncorhynchus mykiss*）和褐鳟（*Salmo trutta*）。

【易感阶段】 在宿主的各生长阶段都可发病，不同基因型的毒株致病力差别较大，死亡率为1%～48%。

【发病水温】 15℃时感染力最强。

【地域分布】 首次报道于苏格兰鲑养殖场。目前，挪威、法国、爱尔兰、英国、意大利和西班牙均有分布，主要流行于欧洲。

【疾病地位】 世界动物卫生组织（OIE）将其列入水生动物疫病名录。

病原

（1）病原为鲑甲病毒（Salmonid alphavirus, SAV）。

（2）属披膜病毒科（*Togaviridae*）、α病毒属（*Alphavirus*）。

（3）病毒粒子呈球状，直径60～70nm，有包膜，有4种衣壳糖蛋白：E1、E2、E3和6K。

（4）基因组为线性单股正链RNA，大小约为12kb，共有6个基因型，包括2个开放阅读框（ORFs）。其中，5′端的ORF编码4个非结构蛋白，依次分别为nsP1、nsP2、nsP3、nsP4；3′端的ORF编码5个结构蛋白，依次分别为衣壳蛋白以及糖蛋白E3、E2、6K和E1。

（5）根据编码E2和nsP3的核酸序列，鲑甲病毒可分为6个基因型（SAV1～SAV6），亚型之间抗原变异性较低，针对特定SAV亚型的单克隆抗体会和其他亚型发生交叉反应。

临床症状和病理学变化

（1）病鱼食欲减退、昏睡、眼突、腹水且粪便拖尾，无法保持在水中的位置，螺旋或绕圈游动，或者在水底不动，但抓捕时会游开，有时出现突然死亡。

（2）解剖可见病鱼心脏苍白，肠道空虚，有黄色黏液，体内脂肪很少，有时在幽门盲肠和四周的脂肪出现出血斑。

（3）胰腺、心肌和骨骼肌病变是主要病症。胰腺组织遭到破坏，胰腺腺泡组织缺失，有时伴随腺泡周围组织纤维化增生，导致病鱼发育不良。

诊断方法

（1）靶器官 所有器官，包括脑、鳃、心脏、胰腺、肾、骨骼肌等。

（2）病毒分离 使用CHSE-214（常用）、BF-2、FHM、SHK-1、EPC、CHH-1细胞系，分离温度为15℃。SAV3和海水SAV2较难观察到CPE。

（3）RT-PCR 能检出SAV的所有基因型，有2组引物可供选择，分别为：

① E2F（5′-CCG-TTG-CGG-CCA-CAC-TGG-ATG-3′）和E2R（5′-CCT-CAT-AGG-TGA-TCG-ACG-GCA-G-3′），扩增产物长度为516bp。

② nsP3F（5′-CGC-AGT-CCA-GCG-TCA-CCT-CAT-C-3′）和nsP3R（5′-TCA-CGT-TGC-CCT-CTG-CGC-CG-3′），扩增长度为490bp。

以上扩增产物经测序后进行判定。

（4）Real-time PCR（定量PCR）

QnsP1F（5′-CCG-GCC-CTG-AAC-CAG-TT-3′）

QnsP1R（5′-GTA-GCC-AAG-TGG-GAG-AAA-GCT-3′）

QnsP1探针（5′-FAM-CTG-GCC-ACC-ACT-TCG-A-MGB-3′）

扩增产物长度为107bp。

（5）酶联免疫吸附试验（ELISA）

防治方法

（1）对养殖场引入的亲鱼或苗种，采取严格的检疫和隔离措施。

（2）选育不带病原的健康苗种。

（3）保持养殖场良好的卫生管理水平，定期进行清塘消毒，杀灭病原体。

（4）内服免疫增强与代谢调节剂，如多糖、多肽、多种维生素。

（5）使用商品化疫苗。

Salmonid alphavirus disease, SAVD

Disease overview

[Disease Characteristic] Also called the Pancreas disease (PD) or Sleeping disease (SD).

[Susceptible Host] Different genotypes has different host. Mainly affect Atlantic salmon (*Salmo salar*), rainbow trout (*Oncorhynchus mykiss*) and brown trout (*Salmo trutta*).

[Susceptible Stage] Clinical disease can occur in all age groups. Virulence varies among different genotypes, with mortality rate ranges from 1% to 48%.

[Outbreak Water Temperature] Highest infectivity at 15℃ .

[Geographic Distribution] The disease was first reported in a fish farm in Scotland. At present, it has been found in Norway, France, Ireland, England, Italy and Spain, and most endemic in Europe.

[Disease Status] OIE-listed Aquatic Animal Disease.

Aetiological agent

(1) Salmonid alphavirus, SAV.

(2) Family: *Togaviridae*. Genus: *Aplhavirus*.

(3) The virus is enveloped and spherical in shape, 60~70nm in diameter. It consists of four capsid glycoproteins: E1, E2, E3 and 6K.

(4) The viral genome is composed of a single-stranded, positive sense RNA containing approximately 12kb, subcategorizing into 6 genotypes. It possesses two Open Reading Frames (ORFs), of which the 5′ end encodes four non-structural proteins, the nsP1, nsP2, nsP3 and nsP4, while the 3′ end of the ORF encodes five structural proteins, the capsid protein and glycoproteins E3, E2, 6K and E1.

(5) According to nucleic acid sequences of E2 and nsP3, SAV can be subcategorized into 6 genotypes (SAV1 ~ SAV6) with low antigenic variation among different genotypes. Monoclonal antibodies against specific SAV genotype will cross-react with other genotype.

Clinical signs and pathological changes

(1) Inappetence, lethargy, exophthalmia, ascites and sticky excretion. Affected fish lose balance, or may swim in spiral or circle, or may sink at the bottom of the pond. Sudden death of fish may occur sometimes.

(2) Upon post-mortem examination, the heart may be pale, and the intestine may be empty with yellowish mucoid discharge with scarce visceral fat. Occasionally ecchymosis is present in the pyloric area, caeca and surrounding adipose tissues.

(3) Lesions in the pancreas, cardiac muscles and bone are the major findings. Pancreatic damage with loss of pancreatic acinar cells, sometimes accompanying with fibrosis, leading to undergrowth of the fish.

Diagnostic methods

(1) Target organs All organs including brain, gill, heart, pancreas, kidney, muscles and bone.

(2) Virus Isolation Use CHSE-214 (most common), BF-2, FHM, SHK-1, EPC and CHH-1 cell lines and incubate at 15℃. CPE is not apparent in SAV3 and marine SAV2.

（3）**Reversed transcriptase-polymerase chain reaction (RT-PCR)** Can detect all SAV genotypes. There are two sets of primers:

① E2F (5′-CCG-TTG-CGG-CCA-CAC-TGG-ATG-3′) and E2R (5′-CCT-CAT-AGG-TGA-TCG-ACG-GCA-G-3′). The size of the amplicon is 516bp.

② nsP3F (5′-CGC-AGT-CCA-GCG-TCA-CCT-CAT-C-3′) and nsP3R (5′-TCA-CGT-TGC-CCT-CTG-CGC-CG-3′). The amplicon size is 490bp.

Perform sequencing on the amplicons for confirmation.

（4）**Real-time PCR (qPCR)** Use the primers QnsP1F (5′-CCG-GCC-CTG-AAC-CAG-TT-3′) and QnsP1R (5′-GTA-GCC-AAG-TGG-GAG-AAA-GCT-3′) with QnsP1 probe (5′-FAM-CTG-GCC-ACC-ACT-TCG-A-MGB-3′). The amplicon size is 107bp.

（5）**Enzyme-linked immunosorbant assay (ELISA)**

Preventative measures

（1）Apply stringent quarantine measures on broodstock, fry and fingerlings introduced to the farm.

（2）Select SPF fry or fingerling for farming.

（3）Maintain good hygiene management in aquaculture farms and regularly disinfection culture ponds to eradicate pathogens.

（4）Oral administration of immunostimulant and metabolic modifiers such as polysaccharides, peptides, and multivitamins.

（5）Use commercial vaccine.

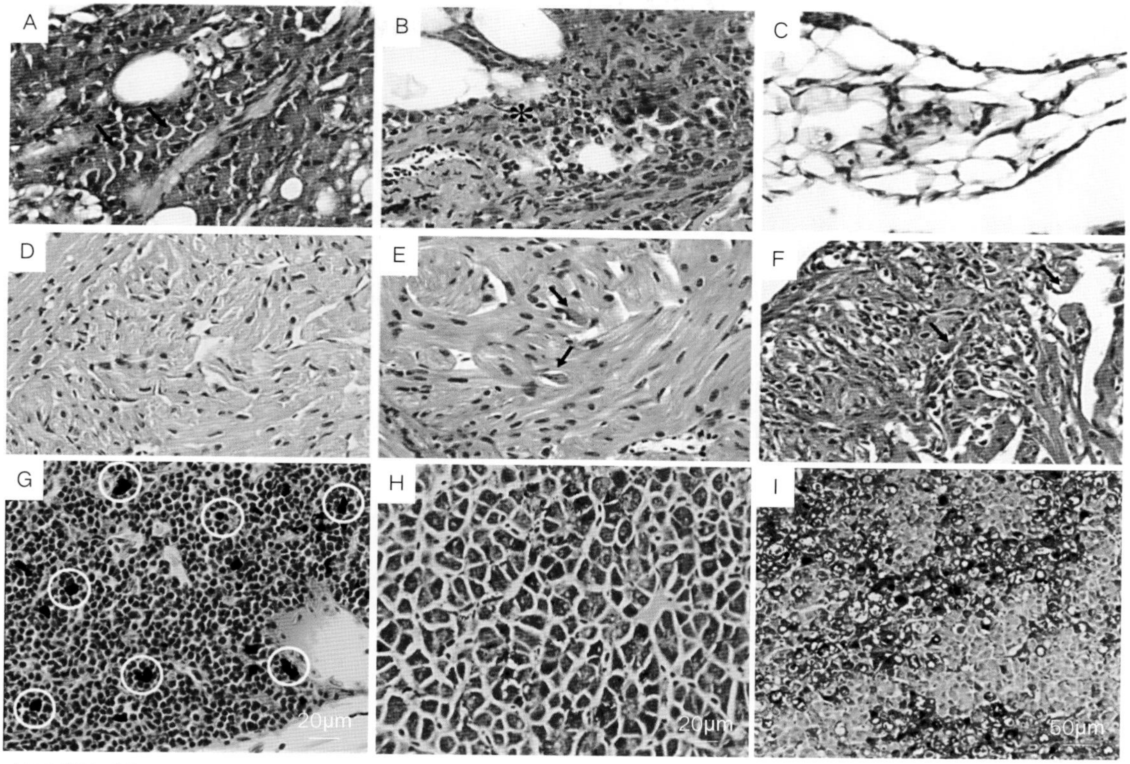

大西洋鲑感染SAV-3后不同细胞及组织的病理变化(HE染色)

A．外分泌胰腺细胞多灶性坏死（箭头） B．胰腺坏死区（星号） C．外分泌组织空泡化 D．正常心肌细胞 E．坏死心肌细胞（箭头） F．心肌细胞坏死（箭头），炎性细胞浸润 G．患病鱼的中肾具有大量黑色素（圆圈） H．严重时，患病鱼苗肝脏中坏死嗜酸性粒细胞增多 I．肝脏细胞质空泡化

[源自Cheng Xu及Tharangani K．Herath]

Histological lesions of affected fish (HE staining)

A．Multifocal necrosis of exocrine pancreatic glands (arrows)

B．Pancreatic necrosis (asterisk) C．Vacuolation of exocrine tissues

D．Normal cardic myofibers E．Necrotic cardiac myofibers (arrows)

F．Cardiac myofiber necrosis with infiltration of inflammatory cells (arrows)

G．Numerous melanomacrophages present in the kidney (circles)

H．Increased number of eosinphils in liver of affected fry I．Vacuolation of hepatocytes

[Source：Cheng Xu and Tharangani K．Herath]

心脏和骨骼肌炎

疾病概述

【概述】 心脏和骨骼肌炎是一种能广泛流行并造成水产养殖业巨大损失的疾病，1999年首次在挪威的养殖大西洋鲑中发现。

【宿主】 主要感染银大麻哈鱼（*Onchorhynchus kisutch*）、大麻哈鱼（*O. keta*）、虹鳟（*O. mykiss*）和马苏大麻哈鱼（*O. masou*）、大西洋鲑（*Salmo salar*）。

【易感阶段】 鱼类洄游后的5 ~ 9个月易感染，死亡率可达20%。网箱养殖的鱼死亡率会更高。

【地域分布】 流行于美国、日本、爱尔兰、挪威和苏格兰等国家。

病原

（1）由一种病毒或几种相关或不相关病毒引起的，主要病原为鱼呼肠孤病毒（Piscine orthoreovirus，PRV）。

（2）属于呼肠孤病毒科（*Reoviridae*），暂时未对其进行分属，被认为与正呼肠孤病毒感染有关。

（3）双链RNA病毒，无包膜，二十面体，直径为70 ~ 80nm。

（4）具有10个基因组片段，系统发育分析通常与基因组片段S1的序列有关。S1序列分为4个基因型（Ⅰ ~ Ⅳ型），从挪威分离的Ⅰ型，分为亚基因Ⅰa和Ⅰb。其中，从智利大西洋鲑中分离出的鱼呼肠孤病毒都为Ⅰb型，而从智利的银大麻哈鱼中分离得到Ⅱ型病毒。基因组长度为23 308bp，由10个双链RNA片段组成。根据其大小，可分为3个大片段（L1、L2、L3），3个中片段（M1、M2、M3）以及4个小片段（S1、S2、S3和S4）。其中，4个小片段分别编码λ蛋白、μ蛋白以及σ蛋白。

临床症状和病理学变化

（1）病鱼食欲不振，游动异常。

（2）出现贫血，鳃苍白，有腹水，心脏发白，肝脏变黄，脾肿大，内脏脂肪周围有出血点。出现黄疸，引起心包炎和心肌炎。

（3）组织切片中可见红细胞中有大量包涵体，心脏组织多灶性或弥漫性损伤。严重时，大多数组织有弥漫性细胞浸润和坏死。

诊断方法

（1）RT-PCR　有2组引物可供选择，分别为：

① PRV-L1 F1（5′-CAC-TCA-CCA-ATG-ACC-CAA-ATG-C-3′）和PRV-L1 R1（5′-TTG-ACA-GTC-TGG-CTA-CTT-CGG-3′），退火温度为54℃，扩增产物长度为937 bp。

② PRV-L1 F2（5′-CTG-AAC-TGC-TAG-TTG-AGG-ATG-G-3′）　和PRV-L1 R2（5′-GCC-AAT-CCA-AAC-AGA-TTA-GG-3′），退火温度为54℃，扩增产物长度为937 bp。

以上扩增产物都需经测序后进行判定。

（2）Real-time RT-PCR　引物分别为：

PRV-F（5′-TCG-TGG-TTC-CAA-TGA-CAG-3′）和PRV-R（5′-CCA-ACC-ACF-AAA-ACC-GAG-3′）。

探针为（5′-6-FAM-ACG-CCT-TAG-AGA-CAA-CAT-GCG-AAG-BHQ-1-3′），退火温度为55℃。

防治方法

（1）保持养殖场良好的卫生管理水平。

（2）对养殖场引入的亲鱼或苗种，采取严格的检疫和隔离措施。

（3）选育不带病原的健康苗种。

（4）内服免疫增强与代谢调节剂，如多糖、多肽、多种维生素。

Heart and skeletal muscle inflammation, HSMI

Disease overview

[Disease Characteristic] Disease causing widespread endemic impact in aquaculture, first reported in cultured Atlantic salmon in Norway in 1999.

[Susceptible Host] Mainly affect silver salmon (*Onchorhynchus kisutchi*), chum salmon (*O. keta*), rainbow trout (*O. mykiss*), masu salmon (*O. masou*) and Atlantic salmon (*Salmo salar*).

[Susceptible Stage] Fish are more susceptible during 5~9 months post-migration, mortality can reach 20%. Mortality may even be higher in net cage culture.

[Geographic Distribution] Endemic in America, Japan, Ireland, Norway and Scotland, etc.

Aetiological agent

(1) Poscine orthoreovirus (PRV). Disease may be caused by infection of one virus or co-infection of various unrelated viruses.

(2) The virus belongs to *Reoviridae*, which has not yet been subcategorized at genus level. It is believed that the disease is related to orthoreovirus.

(3) Double-stranded RNA virus, no capsule, icosahedral with diameter of 70~80nm.

(4) The virus contains 10 genome fragments. Phylogenetic analysis is usually related to the sequence of the genome fragment S1. The viral S1 sequences can be subcategorized into 4 genotypes (Ⅰ ~ Ⅳ). Genotype Ⅰ isolates from Norway can be subdivided into sub-genotype Ⅰa and Ⅰb. All the PRV isolates from Atlantic salmon in Chile are sub-genotype Ⅰb, while PRV isolates from silver salmon in Chile are genotype Ⅱ . The total size of the viral genome is 23, 308bp, composing of 10 dsRNA fragments. These gene fragments can be identified as 3 large fragments (L1, L2 and L3), 3 medium fragments (M1, M2 and M3) and 4 small fragments (S1, S2, S3 and S4) according to the corresponding sizes. Among these gene fragments, the 4 small fragments are responsible for encoding respectively λ , μ and σ proteins.

Clinical signs and pathological changes

(1) Inappetence and abnormal swimming motion.

(2) Anaemia, pale gills, ascites, whitish heart, yellowish liver, enlarged spleen and petechiae on the visceral fat. May cause jaundice, pericarditis and myocarditis.

(3) Histologically, numerous inclusions can be observed in red blood cells. Heart tissues show multifocal or diffuse damage. In severe case, most of the tissues show diffuse inflammatory cells infiltration and necrosis.

Diagnostic methods

(1) Reversed transcriptase-polymerase chain reaction (RT-PCR) There are 2 pairs of primers available.

① PRV-L1 F1 (5′-CAC-TCA-CCA-ATG-ACC-CAA-ATG-C-3′) and PRV-L1 R1 (5′-TTG-ACA-GTC-TGG-CTA-CTT-CGG-3′) with annealing temperature at 54℃. The amplicon size is 937bp.

② PRV-L1 F2 (5′-CTG-AAC-TGC-TAG-TTG-AGG-ATG-G-3′) and PRV-L1 R2 (5′-GCC-AAT-CCA-AAC-AGA-TTA-GG-3′) with annealing temperature at 54 ℃ . The amplicon size is 937bp.

Perform sequencing on the amplicons for confirmation.

(2) Real-time RT-PCR Use the primers PRV-F (5′-TCG-TGG-TTC-CAA-TGA-CAG-3′)

and PRV-R (5′-CCA-ACC-ACT-AAA-ACC-GAG-3′) with probe (5′-6-FAM-ACG-CCT-TAG-AGA-CAA-CAT-GCG-AAG-BHQ-1-3′). The annealing temperature is 55℃ .

Preventative measures

(1) Maintain good hygiene management in aquaculture farms.

(2) Apply stringent quarantine and isolation measures when introducing new broodstock and fingerling to the farm.

(3) Select SPF fry or fingerling for farming.

(4) Oral administration of immunostimulant and metabolic modifiers such as polysaccharides and multivitamins, etc.

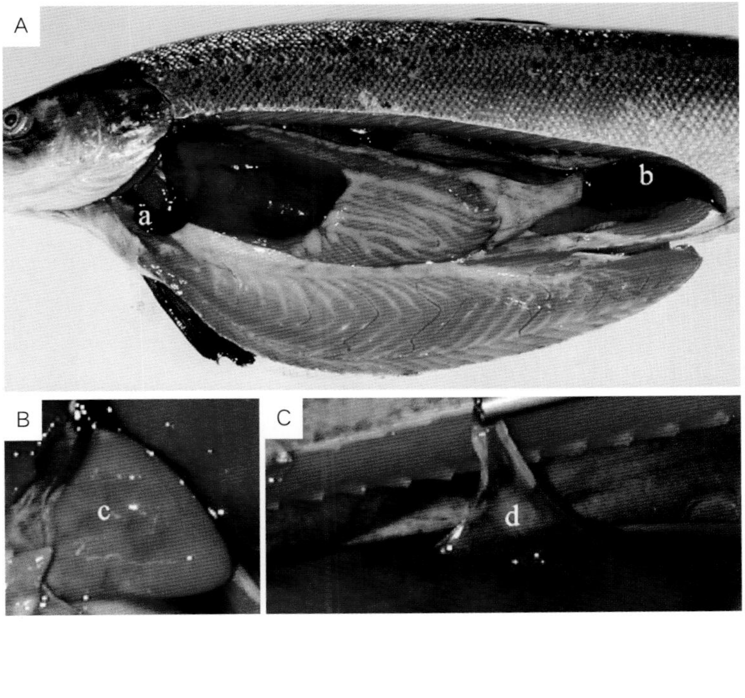

患病大西洋鲑内脏病变

A．此病鱼中可见心脏苍白及心包囊积血，脾脏肿大

B．近观—心脏苍白

C．有些病鱼肝脏出现纤维层

a．心包囊积血 b．脾脏

c．心脏 d．纤维层

[源自T Kongtorp，R & Taksdal，Torunn & Lyngy]

Macroscopic findings in affected *Salmo salar*

A．In this salmon, the heart is pale and there is haemopericardium. The spleen appears swollen B．Closer picture of a pale heart C．A few fish had a fibrinous layer on the liver

a．Haemopericardium b．Spleen

c．Heart d．Fibrinous layer

[Source：T Kongtorp，R & Taksdal，Torunn & Lyngy]

真鲷虹彩病毒病

疾病概述

【概述】 真鲷虹彩病毒病是由真鲷虹彩病毒（RSIV）或传染性脾肾坏死病毒（ISKNV）引起的养殖鱼类高度传染性疾病。

【宿主】 RSIV主要感染真鲷（*Pagrus major*）、黑鲷（*Acanthopagrus schlegeli*）、黄鳍金枪鱼（*Acanthopagrus latus*）、红海鲷（*Evynnis japonica*）、日本鰤（*Seriola quinqueradiata*）、高体鰤（*Seriola dumerili*）、黄体鰤（*Seriola lalandi*）、日本鰤和黄尾鰤的杂交后代（*S. lalandi*×*S. quinquera*）、黄带拟鲹（*Pseudocaranx dentex*）、太平洋蓝鳍金枪鱼（*Thunnus thynnus*）、日本鲅（*Scomberomorus niphonius*）、鲐鲭（*Scomber japonicus*）、竹筴鱼（*Trachurus japonicus*）、日本鹦嘴鱼（*Oplegnathus fasciatus*）、斑石鲷（*Oplegnathus punctatus*）、军曹鱼（*Rachycentron canadum*）、狮鼻鲳鲹（*Trachinotus blochii*）、三线矶鲈（*Parapristipoma trilineatum*）、花尾胡椒鲷（*Plectirhinchus cinctus*）、红鳍裸颊鲷（*Lethrinus haematopterus*）、星斑裸颊鲷（*Lethrinus nebulosus*）、斑鮀（*Girella punctata*）、许氏平鲉（*Sebastes schlegeli*）、大黄鱼（*Pseudosciaena crocea*）、香港鲈（*Epinephelus akaara*）、黑带丽体鲈（*Epinephelus tauvina*）、马拉巴印花鲈（*Epinephelus malabaricus*）、长牙鲈（*Epinephelus bruneus*）、斜带石斑鱼（*Epinephelus coioides*）、青石斑鱼（*Epinephelus awoara*）、巨石斑鱼（*Epinephelus tauvina*）、褐色花纹鲈（*Epinephelus fuscoguttatus*）、日本海鲈（*Lateolabrax japonicas*）、花鲈（*Lateolabrax japonicus*）、尖吻鲈（*Lates calcarifer*）、黑鲈和白鲈杂交鱼（*Morone saxatilis* × *M. chrysops*）、大口黑鲈（*Micropterus salmoides*）、牙鲆（*Paralichthys olivaceus*）、圆斑星鲽（*Verasper variegatus*）、红鳍东方鲀（*Takifugu rubripes*）。ISKNV主要感染鳜（*Siniperca chuatsi*）、中华鲈（*Siniperca chuatsi*）、美国红鱼（*Sciaenops ocellatus*）、鲻鲮（*Mugil cephalus*）、石斑鱼（*Epinephelus* sp.）等。

【易感阶段】 主要危害真鲷幼鱼，对成鱼也有一定影响。

【发病水温】 水温达到25℃以上时容易暴发流行，死亡率差别很大。

【地域分布】 日本、韩国、朝鲜、马来西亚、菲律宾、泰国、新加坡、加拿大、澳大利亚、德国、中国等。

【疾病地位】 世界动物卫生组织（OIE）将其列入水生动物疫病名录。

病原

（1）病原为真鲷虹彩病毒（Red sea bream iridovirual，RSIV）或传染性脾肾坏死病毒（Infection spleen and kidney necrosis virus，ISKNV）。

（2）属虹彩病毒科（*Iridoviridae*）、巨大细胞病毒属（*Megalocytivirus*）。

（3）病毒粒子为正十二面体，直径为140 ～ 160nm，有囊膜。

（4）ISKNV与RSIV的基因序列非常相似。ISKNV的全基因组序列长度为111 362bp，分型以基因型分型为主。根据主要衣壳蛋白基因序列差异性，肿大细胞病毒属可分为3种基因型，基因型差异与病毒地理分布和宿主范围存在密切关系。其中，ISKNV 属基因Ⅱ型。

临床症状和病理学变化

（1）病鱼昏睡，游动减少，不活泼，无活力，呼吸困难。

（2）体色变黑，鳃变灰白色，出现淤斑点。

（3）严重贫血，脾脏肿大。

（4）观察组织切片，可见病鱼的脾出现异常肿大的嗜碱性粒细胞，脾、心、肾、肠、鳃出现异常肿大细胞。

诊断方法

（1）病毒分离 使用石鲈鰭（GF）细胞系，分离温度为25℃。

（2）PCR 分两步进行：

① 1-F（5′-CTC-AAA-CAC-TCT-GGC-TCA-TC-3′）和1-R（5′-GCA-CCA-ACA-CAT-CTC-CTA-TC-3′），退火温度为58℃，扩增产物长度为570bp。若结果为阳性，则判定为RSIV/ISKNV。

② 4-F（5′-CGG-GGG-CAA-TGA-CGA-CTA-CA-3′）和4-R（5′-CCG-CCT-GTG-CCT-TTT-CTG-GA-3′），退火温度为58℃，扩增产物长度为568bp，用于鉴别RSIV和ISKNV。若结果为阳性，则判定为RSIV；若结果为阴性，则判定为ISKNV。

以上扩增产物都需经测序后进行判定。

（3）LAMP 需要3对引物，分别为：

①上游外引物RSIV-F3（5′-CAA-GAA-TGT-CAC-TCA-CCG-CA-3′）和下游外引物RSIV-B3（5′-CAC-CAT-CCA-TCT-CAG-GCA-TG-3′）。

② 上游内引物RSIV-FIP（5′-GCC-AGC-AAA-GGC-AGA-TTC-ACC-TAC-GTG-CAA-AGC-AAT-TAC-ACC-3′），下游内引物GSIV-BIP（5′-TTA-CGA-GAA-CAC-CCC-TCG-GCT-AGG-GGT-CGA-CAG-ATG-TGA-3′）。

③上游环引物RSIV-LF（5′-CCA-CCA-GAT-GGG-AGT-AGA-CTA-C-3′），下游环引物RSIV-LB（5′-TGT-TGA-CAT-ACA-CGG-GAC-TGG-3′），62℃恒温下扩增40min，80℃酶灭活5min。既可在恒温水浴锅中进行，也可在金属浴及各种核酸扩增仪等仪器中进行。

（4）间接免疫荧光试验（IFAT）

防治方法

（1）保持养殖场良好的卫生管理水平。

（2）对养殖场引入的亲鱼或苗种，采取严格的检疫和隔离措施。
（3）选育不带病原的健康苗种。
（4）内服免疫增强与代谢调节剂，如多糖、多肽、多种维生素。
（5）使用商品化疫苗。

Red sea bream iridovirus disease, RSIVD

Disease overview

[Disease Characteristic] Highly infectious disease in fish aquaculture caused by Red sea bream iridovirus (RSIV) or Infectious spleen and kidney necrosis virus (ISKNV).

[Susceptible Host] RSIV mainly affects red sea bream (*Pagrus major*), black progy (*Acanthopagrus schlegeli*), yellow fin sea bream (*Acanthopagrus latus*), crimson sea bream (*Evynnis japonica*), Japanese amberjack (*Seriola quinqueradiata*), greater amberjack (*Seriola dumerili*), yellow tail amberjack (*Seriola lalandi*), hybird of yellowtail amberjack and Japanese amberjack (*S. lalandi×S. quinquera*), striped jack (*Pseudocaranx dentex*), north bluefin tuna (*Thunnus thynnus*), Japanese Spanish mackerel (*Scomberomorus niphonius*), chub mackerel (*Scomber japonicus*), Japanese chub mackerel (*Trachurus japonicus*), Japanese parrot fish (*Oplegnathus fasciatus*), spotted knifekaw (*Oplegnathus punctatus*), cobia (*Rachycentron canadum*), snubnose pompano (*Trachinotus blochii*), chicken grunt (*Parapristipoma trilineatum*), crescent sweetlips (*Plectirhinchus cinctus*), Chinese emperor (*Lethrinus haematopterus*), spangled emperor (*Lethrinus nebulosus*), largescale blackfish (*Girella punctata*), rockfish (*Sebastes schlegeli*), croceine croaker (*Pseudosciaena crocea*), Hong Kong grouper (*Epinephelus akaara*), greasy grouper (*Epinephelus tauvina*), Malabar grouper (*Epinephelus malabaricus*), longtooth grouper (*Epinephelus bruneus*), orange-spotted grouper (*Epinephelus coioides*), yellow grouper (*Epinephelus awoara*), giant grouper (*Epinephelus tauvina*), brown-marbled grouper (*Epinephelus fuscoguttatus*), Japanese sea perch (*Lateolabrax japonicas*), sea bass (*Lateolabrax japonicus.*), barramundi (*Lates calcarifer*), hybrid of striped sea bass and white bass (*Morone saxatilis×M. chrysops*), largemouth bass (*Micropterus salmoides*), bastard halibut (*Paralichthys olivaceus*), spotted halibut (*Verasper variegatus*) and torafugu (*Takifugu rubripes*). ISKNV mainly affects Chinese perch (*Siniperca chuatsi*), red drum (*Sciaenops ocellatus*), flathead mullet (*Mugil cephalus*), and groupers (*Epinephelus* sp.), etc.

[Susceptible Stage] Mainly affect juveniles of red sea bream, but can also affect adult fish.

[Outbreak Water Temperature] Disease outbreaks mostly occur when water temperate is above 25℃. Mortality varies greatly.

[Geographic Distribution] Distributed in Japan, Korea (Rep. of), Korea (DPR), Malaysia, the Philippines, Thailand, Singapore, Canada, Australia, Germany, China, etc.

[Disease Status] OIE-listed Aquatic Animal Disease.

Aetiolgical agent

(1) Red sea bream iridovirus (RSIV), or Infectious spleen and kidney necrosis virus (ISKNV).

(2) Family: *Iridoviridae*. Genus: *Megalocytivirus*.

(3) The virus is dodecahedron in shape, enveloped, and 140~160nm in diameter.

(4) The genetic sequences of ISKNV and RSIV are highly similar. The total length of ISKNV genome is 111, 362bp, subtyping is mainly categorized by genotypes. According to the difference in the protein gene sequence of the main capsule, *Megalocytivirus* can be classified into 3 genotypes. The difference among these genotypes is closely related to the geographical distribution of the virus and its susceptible host range. ISKNV belongs to genotype Ⅱ.

Clinical signs and pathological changes

(1) Lethargic, reduced swimming, inactive and dyspnoea.

(2) Darkened body color. Gills appear greyish-white with petechiae.

(3) Severe anaemia, spleen enlargement.

(4) Histologically, abnormally large basophils is present in the spleen; swollen cells can be observed in the spleen, heart, kidney, intestine and gill.

Diagnostic methods

(1) **Virus isolation** Use Grunt fin (GF) cell line to incubate at 25℃.

(2) **Polymerase chain reaction (PCR)** 2 steps are involved.

① First step primers: 1-F (5′-CTC-AAA-CAC-TCT-GGC-TCA-TC-3′) and 1-R (5′-GCA-CCA-ACA-CAT-CTC-CTA-TC-3′) with annealing temperature at 58℃. The amplicon size is 570bp. If the result is positive, then it is RSIV or ISKNV.

② Second step primers: 4-F (5′-CGG-GGG-CAA-TGA-CGA-CTA-CA-3′) and 4-R (5′-CCG-CCT-GTG-CCT-TTT-CTG-GA-3′) with annealing temperature at 58℃. The amplicon size is 568bp. The second step is used to differentiate RSIV and ISKNV, indicating RSIV if the result is positive, and ISKNV if the result is negative.

Perform sequencing on the amplicons for confirmation.

(3) **Loop-mediated isothermal amplification (LAMP)** There are 3 pairs of primers available.

① External forward primer RSIV-F3 (5′-CAA-GAA-TGT-CAC-TCA-CCG-CA-3′) and external reverse primer RSIV-B3(5′-CAC-CAT-CCA-TCT-CAG-GCA-TG-3′).

② Internal forward primer RSIV-FIP (5′-GCC-AGC-AAA-GGC-AGA-TTC-ACC-TAC-GTG-CAA-AGC-AAT-TAC-ACC-3′) and internal reverse primer GSIV-BIP (5′-TTA-CGA-GAA-CAC-CCC-TCG-GCT-AGG-GGT-CGA-CAG-ATG-TGA-3′).

③ Forward loop primer RSIV-LF(5′-CCA-CCA-GAT-GGG-AGT-AGA-CTA-C-3′) and reverse loop primer RSIV-LB(5′-TGT-TGA-CAT-ACA-CGG-GAC-TGG-3′). Carry out amplification for 40 minutes at 62℃ followed by enzymatic inactivation for 5 minutes at 80℃ .

These procedures can be carried out in a water bath, metal bath or in any other kinds of nucleic acid amplification instrument.

(4) Immunofluoresence antibody test (IFAT)

Preventative measures

(1) Maintain good hygiene standards in the farm.

(2) Apply stringent quarantine measures on broodstock, fry and fingerling introduced to the farm.

(3) Select SPF fry and fingerling for farming.

(4) Oral administration of immunostimulant and metabolic modifiers, such as polysaccharides, polypeptide, multivitamins, etc.

(5) Use commercial vaccine.

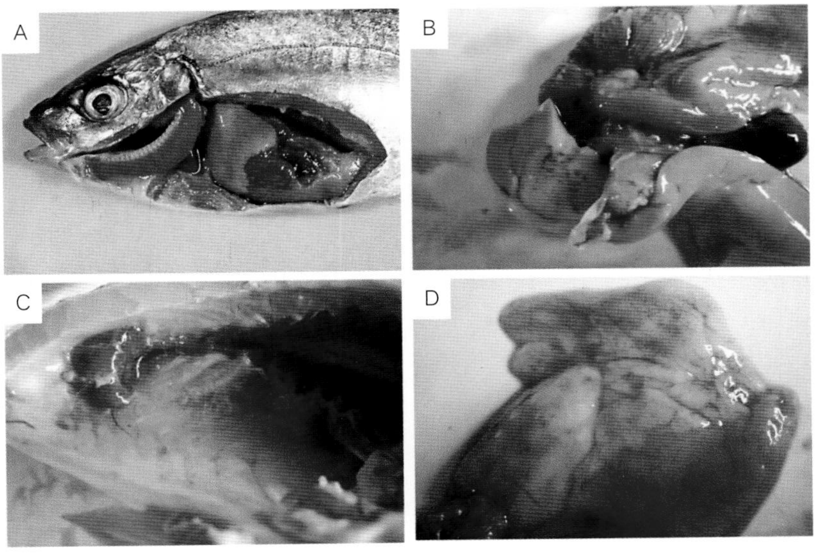

患RSIV病鱼

A. 鳃、肝脏的褪色　B. 肝脏出血　C. 肾脏水肿　D. 内脏脂肪组织出血

[源自《新鱼病图鉴》，小川和夫；杨慧英]

Macroscopic findings of affected fish

A. Discoloration of the gill and liver　B. petechial haemorrhage in the liver　C. Renal oedema　D. Haemorrhage of visceral adipose tissues

[Source: *New Atlas of Fish Diseases*, Kazuo Ogawa; Huiying Yang]

淋巴囊肿病

疾病概述

【概述】 淋巴囊肿病是一种慢性病毒性疾病。

【宿主】 可以感染123种以上温水域和低温水域海水、淡水鱼类，包括慈鲷科（Cichlidae）、丝足鲈科（Osphronemidae）、鲈科（Centrarchidae）、虾虎鱼科（Gobiidae）、蝴蝶鱼科（Chaeto dontidae）、雀鲷类（Pomacentridae）、石首鱼科（Sciaenidae）、鮨科（Serranidae）和鲽科（Pleuronectidae）。

【易感阶段】 皮肤有外伤的鱼易发病。

【发病水温】 一般在15 ～ 20℃暴发，高水温季节也可发病，感染率可高达80%，死亡率可达30%。

【地域分布】 该病广泛见于世界各地的淡水、半咸水和海水中，主要发生于欧洲、美洲、亚洲。近几年来，广东、浙江、山东和河北等省份均有报道。

病原

（1）病原为淋巴囊肿病毒（Lymphocystis disease virus，LCDV）。

（2）属虹彩病毒科（*Iridoviridae*）、淋巴囊肿病毒属（*Lymphocystivirus*）。

（3）为双链DNA病毒，呈二十面体，直径一般介于130 ～ 330nm。病毒粒子因宿主、环境不同，大小可发生一定变化。

（4）衣壳由2层组成，外层可被蛋白酶K消化，内层可被磷脂酶A2消化。化学成分包括1.6%的DNA、42.3%的蛋白质、17.5%的脂类以及39%的未命名成分。其中，脂类大部分为磷脂，未命名成分可能主要为糖类。

（5）鱼类淋巴囊肿病毒至少存在7种基因型：基因型Ⅰ（包括LCDV-1）、基因型Ⅱ（包括牙鲆分离株LCDV-jf、LCDV-cn和LCDV-C等）、基因型Ⅲ（包括岩鱼分离株LCDV-rf）、基因型Ⅳ（包括军曹鱼分离株LCDV-rc和鲈分离株LCDV-sb）、基因型Ⅴ（包括玻璃拉拉鱼分离株LCDV-cb）、基因型Ⅵ（珍珠鱼分离株LCDV-tl）、基因型Ⅶ（金头鲷分离株LCDV-sa）。

临床症状和病理学变化

（1）病鱼的皮肤和鳍部可见乳白色或黑白色的结节，有时单一存在，有时呈团状。

（2）肝、脾、肾、心脏和肠黏膜下层均有不同程度的变性、坏死和炎症反应。症状严重的病鱼内脏表面可见由囊肿细胞组成的白色小颗粒。

（3）成纤维细胞呈现肥大，同时，在其肥大细胞的细胞质内具有嗜碱性的包涵体，成熟的淋巴囊肿细胞外具有一层透明的囊膜，被感染的成纤维细胞很快生长成直径约1mm的淋巴囊肿细胞。

诊断方法

（1）酶联免疫吸附试验（ELISA）

（2）PCR　引物为：F（5′-CAG-GTA-CAA-ACA-GCA-CCT-AAA-CAT-G-3′）和R（5′-CAC-CGR-CAA-AGA-TTA-CAG-GAG-AAG-3′），退火温度为50.5℃，扩增产物长度为172bp。扩增产物经测序后判定。

防治方法

（1）保持养殖场良好的卫生管理水平。

（2）对养殖场引入的亲鱼或苗种，采取严格的检疫和隔离措施。

（3）把好饲料关，少用或不用鲜活饵料，考虑对可能受病毒污染的饵料进行消毒处理。

（4）选育不带病原的健康苗种。

（5）疾病暴发后，要对病鱼实行集中消毒或销毁，养殖废水须经过消毒处理后才可排放。

（6）22 ~ 25℃高温结合过氧化氢药浴，可使病鱼的体表囊肿消失。

Lymphocystis disease, LCD

Disease overview

[Disease Characteristic] It is a chronic viral disease.

[Susceptible Host] Over 123 species of temperate or cool marine or freshwater fishes, including families of Cichlidae, Osphronemidae, Centrarchidae, Gobiidae, Chaetodontidae, Pomacentridae, Sciaenidae, Serranidae and Pleuronectidae.

[Susceptible Stage] Fish with external injuries are prone to develop clinical disease.

[Outbreak Water Temperature] Disease outbreaks generally occur at 15~20℃ , though it can also occur in warm water temperature seasons. Mortality and morbidity rates can reach as high as 30% and 80%, respectively.

[Geographic Distribution] Widely distributed in freshwater, seawater and brackish water worldwide. Disease outbreaks mostly occur in Europe, America, Asia, and have also been reported in different provinces in China, such as Guangdong, Zhejiang, Shandong and Hebei in recent years.

Aetiological agent

(1) Lymphocystis disease virus, LCDV.

(2) Family: *Iridoviridae*. Genus: *Lymphocystivirus*.

(3) The virus consists of a double-stranded DNA, and is symmetrically icosahedrons in shape, 130~330nm in diameter, of which the size of the virions can vary according to different hosts and environments.

(4) The capsid is composed of two layers. The external layer can be digested by proteinase K while the internal layer can be digested by phospholipase A2. Chemical component includes 1.6% DNA, 42.3% protein, 17.5% lipids, in which majority of the lipid content is lecithin, and 39% unnamed components. Among them, most of the lipids are phospholipids, and the unnamed components may be mainly sugars.

(5) There are at least 7 genotypes of LCDV: genotype Ⅰ (LCDV-1), genotype Ⅱ (including gingival isolate LCDV-jf, LCDV-cn and LCDV-C,etc), genotype Ⅲ (including rock fish isolate LCDV-rf), genotype Ⅳ (including cobia isolate LCDV-rc and striped bass isolate LCDV-sb), genotype Ⅴ (including glass zipper isolate LCDV-cb) and genotype Ⅵ (including pearl fish isolate LCDV-t1) and genotype Ⅶ (including gilt-head bream isolate LCDV-sa).

Clinical signs and pathological changes

(1) Single or clusters of milky white or black and white nodules on the skin and fins.

(2) Various degrees of degeneration, necrosis and inflammation in the liver, spleen, kidney, heart and submucosa of intestines. In severe cases, small white cysts can be seen on the surface of the viscera.

(3) Fibroblasts appear hypertrophic, with basophilic intracytoplasmic inclusion bodies in mast cells. Mature lymphocytic cyst is circumscribed by a layer of transparent capsule. Infected fibroblasts will rapidly grow and form lymphocytic cysts of around 1mm in diameter.

Diagnostic methods

(1) Enzyme linked immunosorbent assay (ELISA)

(2) Polymerase chain reaction (PCR) Use forward primer F (5′-CAG-GTA-CAA-ACA-GCA-CCT-AAA-CAT-G-3′) and reverse primer R (5′-CAC-CGR-CAA-AGA-TTA-CAG-GAG-AAG-3′) with annealing temperature at 50.5℃. The amplicon size is 172bp.Perform sequencing on the amplicon for confirmation.

Preventative measures

(1) Maintain good hygiene management in aquaculture farm.

(2) Apply stringent quarantine measures on broodstock, fry and fingerling introduced to the farm.

(3) Maintain good feed management and reduce the use of live feed. Consider sterilization of potentially contaminated feed.

(4) Select SPF fry or fingerling for farming.

(5) After outbreak of disease, the affected fish must be sterilized or eliminated. Waste water from the farm must be disinfected before release.

(6) Warm water bath at 22~25℃ with hydrogen peroxide can treat the cystic lesions on the fish body.

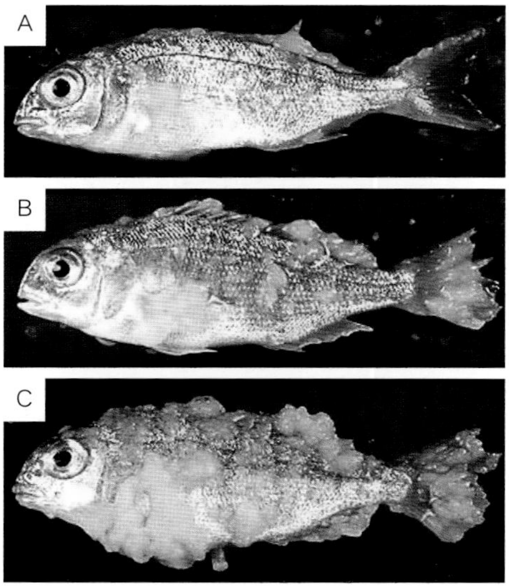

幼鲷淋巴囊肿病的临床特征

A. 感染初期　B. 中度感染　C. 严重感染

[源自 KVITT et al.,2008]

Clinical features of Lymphocystis disease in young sea bream

A. Mild infection at initial stage　B. Medium intensity infection

C. Severe infection

[Source: KVITT et al.,2008]

病毒性神经坏死病

疾病概述

【概述】 病毒性神经坏死病又称病毒性脑病和视网膜病（Viral encephalopathy and retinopathy，VER），是一种分布广泛的海水鱼的严重疫病。

【宿主】 牙鲆（*Paralichthys olivaceus*）、大菱鲆（*Scophthalmus maximus*）、红鳍东方鲀（*Takifugu rubripes*）、尖吻鲈（*Lates calcarifer*）、欧洲舌齿鲈（*Dicentrarchus labrax*）、赤点石斑鱼（*Epinephelus akaara*）、棕点石斑鱼（*E. fuscogutatus*）、玛拉石斑鱼（*E. malabaricus*）、蜂巢石斑鱼（*E. merra*）、七带石斑鱼（*E. septemfasciatus*）、巨石斑鱼（*E. tauvina*）、黄带拟鲹（*Pseudocaranx dentex*）、条石鲷（*Oplegnathus fasciatus*）、条斑星鲽（*Verasper moseri*）、庸鲽（*Hippoglossus hippoglossus*）等近60种鱼类。

【易感阶段】 鱼苗及幼鱼皆会感染，感染程度以孵出十多天的水花（0.8 ~ 1.2cm）最为严重，发病快且死亡率高；体长3 ~ 5cm的鱼苗感染程度差异较大，有严重感染、中度感染与轻微感染，发病过程较水花长，死亡的速度较慢，但死亡率仍高；体长7 ~ 9cm的幼鱼皆为轻微至中度感染，发病及死亡率较低。

【发病水温】 不同基因型发病水温不同。SJNNV型（拟鲹神经坏死病毒型）和RGNNV型（石斑鱼神经坏死病毒型）的发病水温为18 ~ 26℃。在阴天，溶解氧低于4mg/L时，发病鱼和死亡鱼数量明显增加。水温上升到28℃以上，病情有所缓解，死亡数明显减少。而BFNNV型（鲽神经坏死病毒型）主要感染冷水鱼类，可在6℃感染庸鲽。

【地域分布】 发生于中国、印度、印度尼西亚、伊朗、日本、韩国、马来西亚、菲律宾、泰国、越南、澳大利亚、法国、希腊、以色列、意大利、马耳他、葡萄牙、西班牙、突尼斯、英国、挪威、加勒比地区、加拿大和美国。

病原

（1）病原为神经坏死病毒（Nervous necrosis virus，NNV）。

（2）属野田村病毒科（*Nodaviridae*）、β 野田村病毒属（*β-Nodavirus*）。RNA病毒，为最小的动物病毒。

（3）病毒粒子呈球形，二十面体，直径约25nm，无囊膜，类晶格状或单个或成团状排列在细胞质内。

（4）基因组由2个分子的单股正链RNA组成：RNA1（3.1kb）编码复制酶（110 ku），RNA2（1.4kb）编码衣壳蛋白（42ku）。

（5）根据RNA2的T4可变区域的分子进化研究，可将神经坏死病毒分为4个主要基因型：SJNNV型（拟鲹神经坏死病毒型）、TPNNV型（鲀神经坏死病毒型）、BFNNV型（鲽神经坏死病毒型）和RGNNV型（石斑鱼神经坏死病毒型）。

临床症状和病理学变化

（1）病鱼呈现多种异常游动行为，如螺旋式、涡旋状或腹部朝上（有时伴有鱼鳔发炎）或卧于池塘底部、快速打转或向前移动，失去食欲。病鲽症状不太明显，可能会滞留在池塘底部，身体蜷曲，头和尾向上翘。

（2）体表和鳃无明显的外部症状，解剖后可见鳔明显膨胀。

（3）组织病理观察，可见患病鱼中枢神经组织空泡变性，通常在视网膜中心层出现空泡，多数种类的鱼都会出现神经性坏死。小鱼苗损伤更严重，较大鱼的损伤主要出现在视网膜上。

诊断方法

（1）病毒分离 使用SSN-1细胞系、GF-1细胞系或E-11细胞系，SJNNV、TPNNV、BFNNV的分离温度为20℃，RGNNV的分离温度为25℃。

（2）RT-PCR 有3组引物可供选择，分别为：

① Q-RdRP-1（5′-GTG-TCC-GGA-GAG-GTT-AAG-GAT-G-3′）和Q-RdRP-2（5′-CTT-GAA-TTG-ATC-AAC-GGT-GAA-CA-3′），退火温度为58℃，扩增产物长度为273bp。

② Q-CP-1（5′-CAA-CTG-ACA-ACG-ATC-ACA-CCT-TC-3′）和Q-CP-2（5′-CAA-TCG-AAC-ACT-CCA-GCG-ACA-3′），退火温度为58℃，扩增产物长度为230bp。

③ F2（5′-CGT-GTC-AGT-CAT-GTG-TCG-CT-3′）和R3（5′-CGA-GTC-AAC-ACG-GGT-GAA-GA-3′），退火温度为55℃，扩增产物长度约430bp。

扩增产物经测序后判定。

（3）Real-time RT-PCR（RT-qPCR） 有3组引物可供选择：

① P1（5′-GGT-ATG-TCG-AGA-ATC-GCC-C-3′）和P2（5′-TAA-CCA-CCG-CCC-GTG-TT-3′），探针（5′-5FAM-TTA-TCC-CAG-CTG-GCA-CCG-GC-BHQ1-3′），退火温度为58℃，扩增产物长度为194bp。

② Qr2TF（5′-CTT-CCT-GCC-TGA-TCC-AAC-TG-3′）和qR2TR（5′-GTT-CTG-CTT-TCC-CAC-CAT-TTG-3′），探针R2（5′-6FAM-CAA-CGA-CTG-CAC-CAC-GAG-TTG-BHQ1-3′），退火温度为58℃，扩增产物长度为93bp。

③ RNA2 F（5′-CAA-CTG-ACA-RCG-AHC-ACA-C-3′）和RNA2 R（5′-CCC-ACC-AYT-TGG-CVA-C-3′），探针RNA2（5′-6FAM-TYC-ARG-CRA-CTC-GTG-GTG-CVG-BHQ1-3′），退火温度为58℃，扩增产物长度为69bp。

防治方法

（1）加强饲料管理，避免用生鲜鱼作为饵料。

（2）可采用紫外线消毒孵化场的进水，设置安全屏障，定期清塘，消毒水箱和生物滤膜，对设施和工具进行消毒。

（3）用臭氧处理过的海水清洗受精卵，或用臭氧/氯制剂处理养殖用水。

（4）强化亲本培育，减少应激因素，如给亲本投喂足量饵料、降低鱼苗和幼鱼的放养密度等。

（5）对亲鱼性腺进行PCR检测，选择阴性个体进行繁育。

Viral nervous necrosis, VNN

Disease overview

[Disease Characteristic] An important widely distributed marine fish disease, also named as Viral Encephalopathy and Retinopathy (VER).

[Susceptible Host] Affect almost 60 fish species including olive flounder (*Paralichthys olivaceus*), turbot (*Scophthalmus maximus*), Japanese puffer (*Takifugu rubripes*), barramundi (*Lates calcarifer*), European bass (*Dicentrarchus labrax*), Hong Kong grouper (*Epinephelus akaara*), brown marbled grouper (*E. fuscogutatus*), Malabar grouper (*E. malabaricus*), honeycomb grouper (*E. merra*), sevenband grouper (*E. septemfasciatus*), greasy grouper (*E. tauvina*), striped jack (*Pseudocaranx dentex*), barred knifejaw (*Oplegnathus fasciatus*), barfin flounder (*Verasper moseri*), Alantic halibut (*Hippoglossus hippoglossus*), etc.

[Susceptible Stage] Fingerlings and juveniles can be affected, causing highest severity in fry of around 10 days old (0.8~1.2cm) with rapid disease progress and high mortality rate. Disease situation varies greatly in fingerlings of 3~5cm from severe, moderate to mild. The disease progress in these fingerlings is longer, and the death speed is slower than those of fry, but the mortality rate is still high. Juveniles of 7~9cm show mild to moderate disease with lower morbidity and mortality rates.

[Outbreak Water Temperature] Outbreak water temperature is various based on different genotype. For SJNNV (striped jack nervous necrosis virus) and RGNNV (red-spotted grouper nervous necrosis virus), disease outbreaks occur at around 18~26℃ . On cloudy days, the number of diseased fish and dead fish markedly increase when dissolved oxygen level is below 4mg/L. When water temperature is above 28℃ , the outbreak starts to resolve with marked decrease of mortality. However, BFNNV (barfin flounder nervous necrosis virus) is mainly infect cold water fish, and it can even infect *Hippoglossus hippoglossus* at 6℃ .

[Geographical Distribution] The disease occurs in China, India, Indonesia, Iran, Japan, Korea, Malaysia, the Philippines, Thailand, Vietnam, Australia, France, Greece, Israel, Italy, Malta,

Portugal, Tunisia, England, Norway, Caribbean, Canada and USA.

Aetiological agent

(1) Nervous necrosis virus, NNV.

(2) Family: *Nodaviridae*. Genus: *Beta nodavirus*. The smallest RNA virus of animals.

(3) The virion is spherical in shape, icosahedral, 25nm in diameter and non-enveloped, found as lattice-like structure, in single or in cluster inside the cytoplasm.

(4) The genome is composed of two molecules of positive-sense single stranded RNA: RNA1 (3.1kb) encoding replicase (110ku) and RNA2 (1.4kb) encoding capsid protein (42ku).

(5) According to phylogenetic analysis of the T4 variable region of RNA2, four major genotypes can be identified: SJNNV (Striped jack nervous necrosis virus), TPNNV (Tiger puffer nervous necrosis virus), BFNNV (Barfin flounder nervous necrosis virus), and RGNNV (Red-spotted grouper nervous necrosis virus).

Clinical signs and pathological changes

(1) Affected fish show various abnormal swimming motions such as spiraling, whirling, upside-down swimming (sometimes accompanied with swimming bladder inflammation) or stay at the bottom of the pond, and with abnormal rapid swimming. Inappetence. Affected halibuts usually show unapparent clinical signs, may stay at the bottom of the pond in an opisthotonic posture.

(2) No apparent lesions on the body surface and gill, overinflation of the swim bladder is present in post-mortem examination.

(3) Histopathological findings include vacuolation of the central nervous tissues of clinically affected fish, usually at the central retina, and necrosis of nervous tissues in most fish species. The lesions are most severe in fry, while the lesions are mainly restricted to the retina in older fish.

Diagnostic methods

(1) **Virus isolation** Use SSN-1, GF-1 or E-11 cell lines and inoculate at 20℃ for SJNNV, TPNNV and BFNNV; and 25℃ for RGNNV.

(2) **Reverse transcriptase-polymerase chain reaction (RT-PCR)** Three pairs of primers are available.

① Q-RdRP-1 (5′-GTG-TCC-GGA-GAG-GTT-AAG-GAT-G-3′) and Q-RdRP-2 (5′-CTT-GAA-TTG-ATC-AAC-GGT-GAA-CA-3′) with annealing temperature at 58℃. The amplicon size is 273bp.

② Q-CP-1 (5′-CAA-CTG-ACA-ACG-ATC-ACA-CCT-TC-3′) and Q-CP-2 (5′-CAA-TCG-AAC-ACT-CCA-GCG-ACA-3′) with annealing temperature at 58℃. The amplicon size is 230bp.

③ F2 (5′-CGT-GTC-AGT-CAT-GTG-TCG-CT-3′) and R3 (5′-CGA-GTC-AAC-ACG-GGT-

GAA-GA-3′) with annealing temperature at 55℃. The amplicon size is 430bp.

Perform sequencing on the amplicons for confirmation.

(3) **Real-time RT-PCR (RT-qPCR)** Three sets of primers and probes are available.

① P1 (5′-GGT-ATG-TCG-AGA-ATC-GCC-C-3′) and P2 (5′-TAA-CCA-CCG-CCC-GTG-TT-3′) with probe (5′-5FAM-TTA-TCC-CAG-CTG-GCA-CCG-GC-BHQ1-3′). The annealing temperature is 58℃ and the amplicon size is 194bp.

② Qr2TF (5′-CTT-CCT-GCC-TGA-TCC-AAC-TG-3′) and qR2TR (5′-GTT-CTG-CTT-TCC-CAC-CAT-TTG-3′) with R2 probe 2 (5′-6FAM-CAA-CGA-CTG-CAC-CAC-GAG-TTG-BHQ1-3′). The annealing temperature is 58℃ and the amplicon size is 93bp.

③ RNA2 F (5′-CAA-CTG-ACA-RCG-AHC-ACA-C-3′) and RNA2 R (5′-CCC-ACC-AYT-TGG-CVA-C-3′) with RNA2 probe (5′-6FAM-TYC-ARG-CRA-CTC-GTG-GTG-CVG-BHQ1-3′). The annealing temperature is 58℃ and the amplicon size is 69bp.

Preventative measures

(1) Strengthen feed management and avoid using live fish as bait.

(2) Strengthen general hygiene measures: sterilize water going to the hatcheries by UV light. Set up safety screens, stamp out the ponds regularly, sterilize the water tank and biofilters, disinfect the facilities and equipment.

(3) Sterilize fertilized eggs with ozone-treated seawater, or treat culture water with ozone or chlorinated agents.

(4) Reduce stress factors and enhance spawning, such as providing sufficient amount of feed to breeding fish and reducing farming density of fry and fingerlings.

(5) Perform PCR test on broodstock and select those with negative results for breeding.

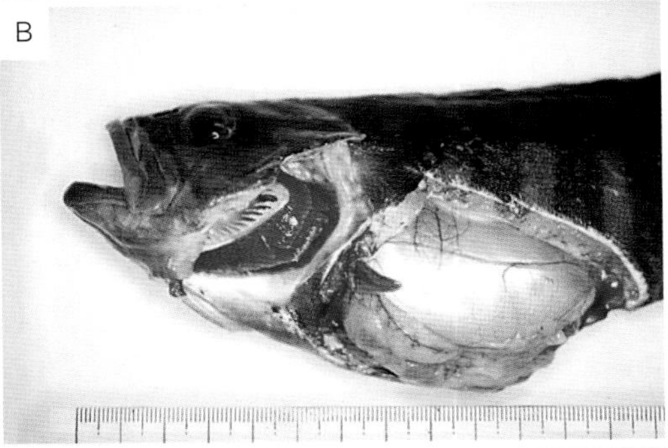

患VNN病鱼临床症状及病变

A．病鱼以翻转状态在水面游动

B．行翻转游动的患病鱼出现鳔扩张症状

C．正常与患病鱼苗的体色对比

[源自《新鱼病图鉴》，小川和夫；利洋]

Clinical signs and macroscopic findings of affected fish

A．Affected fish showing upside-down swimming near the water surface of the net cage

B．Marked distension of the swimming bladder

C．Darkening of body color of affected fish fry (top) comparing with normal fish fry (bottom)

[Source：*New Atlas of Fish Diseases*, Kazuo Ogawa；LIYANG AQUATIC]

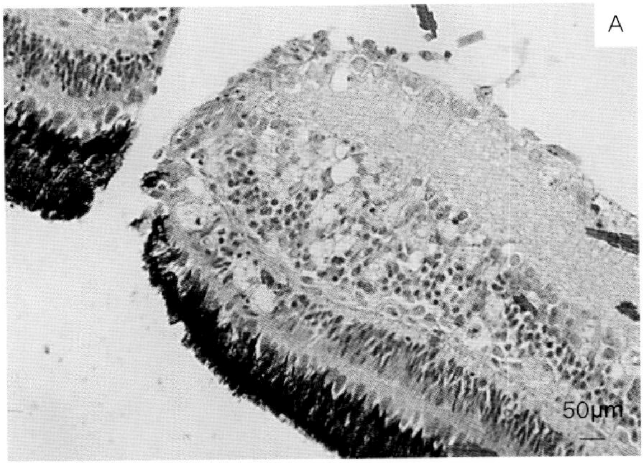

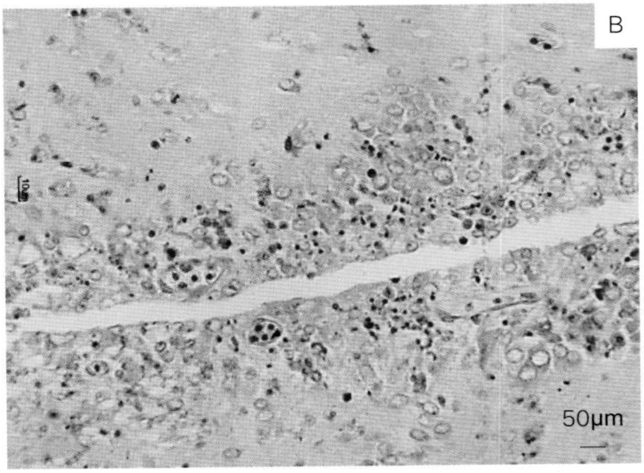

病鱼细胞及组织病变（HE染色）

A．脑组织轻微瘀血，细胞散在坏死，局部组织坏死较重

B．脑组织明显空泡化

[源自利洋]

Histological lesions of affected fish. (HE staining)

A．Mild congestion in nervous tissue, sporadic focal necrosis of nervous cells, and more severe necrosis of local tissue

B．Marked vacuolation of nervous cells

[Source：LIYANG AQUATIC]

传染性胰脏坏死病

疾病概述

【概述】 传染性胰脏坏死病是一种以鲑科鱼类内脏实质器官出血、坏死为主要病理特征的高度接触性、急性传染病。

【易感宿主】 宿主范围广，易感染大西洋鲑（*Salmo salar*）、虹鳟（*Oncorhynchus mykiss*）、褐鳟（*Salmo trutta fario* Linnaeus）、远东红点鲑（*Salvelinus leucomaenis*）和几种太平洋大麻哈鱼类，可感染欧洲舌齿鲈（*Dicentrarchus labrax*）、五条鰤（*Seriola quinqueradiata*）、大菱鲆（*Scophthalmus maximus*）、欧洲黄盖鲽（*Limanda limanda*）、庸鲽（*Hippoglossus hippoglossus*）、鳎（*Solea solea*）、塞内加尔鳎（*Solea senegalensis*）、大西洋鳕（*Gadus morhua*）。

【易感阶段】 小于6个月的幼鱼受害最为严重，2 ~ 10周龄的虹鳟鱼苗，在水温10 ~ 12℃时，感染率和死亡率可高达20% ~ 100%。20周龄以后的鱼种一般不发病，但可成为终身带毒者。

【发病水温】 发病水温一般为10 ~ 15℃。成年鲑科鱼类及鲤(*Cyprinus carpio*)、狗鱼(*Esox reicherti*)等非鲑科鱼类可为亚临床感染的病毒携带者。

【地域分布】 广泛流行于欧洲、亚洲、北美洲、南美洲和非洲南部各国。

病原

（1）病原为传染性胰脏坏死病毒（Infectious pancreatic necrosis virus，IPNV）。

（2）属双核糖核酸病毒科（*Binaviridae*）、液体双核糖核酸病毒属（*Aquabirnavirus*）。

（3）病毒呈球状，无囊膜，直径为50 ~ 75nm，二十面体对称，有1层外壳。

（4）基因组含A/B 2个线状双股RNA分子，大小为6kb。A节段有2个开放阅读框，较小的阅读框编码分子质量为16 500u的非结构蛋白VP5；较大的阅读框编码1个聚合蛋白加工后，形成VP2、VP3及VP4。B节段为2.8kb，编码RNA聚合酶VP1，分子质量94ku，以基因组结合蛋白（VPg）的形式存在，将A、B2个节段的末端紧紧相连。

（5）含有10个血清型，包括A组血清型A1（West Buxton，WB）、A2（Spajarup，Sp）、A3（Abild，Ab）、A4（Hecht，He）、A5（Tellina，Te）、A6（Canada3，C3）和A9（Jasper，Ja），B组血清型B1（Tellinavirus，TV-1）。血清学A1主要出现在美国，血清型A6 ~ A9主要出现在加拿大，血清型A2 ~ A5和B1多出现在欧洲和亚洲。

临床症状和病理学变化

（1）排出线状黏液便，游动失调，常作垂直回转运动，直至死亡。

（2）体色变黑，眼球突出，腹部膨胀，体腔内有腹水。

（3）食欲丧失，解剖后可见消化道内无食物，肠内见有硬黄色物，或灰白卡他性渗出物。胃幽门盲囊及体前部脂肪组织、生殖器和内脏器官有出血点。

（4）组织病理观察可见，胰腺广泛性坏死，肾脏也可见坏死现象。

诊断方法

（1）病毒分离 使用蓝鳃鱼细胞系（BF-2）、鲤上皮瘤细胞系（EPC）、虹鳟性腺细胞系（RTG-2）、胖头鲂肌细胞系（FHM）、大鳞大麻哈鱼胚胎上皮细胞系（CHSE-214），分离温度为15 ℃。

（2）RT-PCR 引物为：IPNV-F（5′-CCA-GCG-AAT-ATT-TTC-TCC-ACC-A-3′）和IPNV-R（5′-AGG-AGA-TGA-CAT-GTG-CTA-CAC-CG-3′），退火温度为55℃，扩增产物长度224 bp。

扩增产物测序后进行进一步诊断。

（3）中和试验 将细胞培养物做系列稀释，分别加入抗IPNV参考血清、细胞培养液，20℃培养，计算中和指数，中和指数大于50，则可判定为疑似传染性胰脏坏死病毒感染。

（4）酶联免疫吸附试验（ELISA）

防治方法

（1）保持养殖场良好的卫生管理水平。

（2）对养殖场引入的亲鱼或苗种，采取严格的检疫和隔离措施。

（3）选育不带病原的健康苗种。

（4）每立方米水体用50g碘伏（PVP-I）消毒鱼卵15min。

（5）疾病早期用PVP-I拌饲投喂，每千克鱼体重每天用有效碘1.64 ～ 1.91g，连续投喂15d。

（6）内服免疫增强与代谢调节剂，如多糖、多肽、多种维生素。

（7）使用商品化疫苗。

Infectious pancreatic necrosis, IPN

Disease overview

[Disease Characteristic] An highly contagious and acute infectious disease of salmonids. It causes haemorrhage and necrosis of visceral organs.

[Susceptible Host] Wide host range. Atlantic salmon (*Salmo salar*), rainbow trout

(*Oncorhynchus mykiss*), brown trout (*Salmo trutta fario* Linnaeus), whitespotted char (*Salvelinus leucomaenis)* and several other Pacific chum salmons species are easily susceptible. The virus can also infect European bass (*Dicentrarchus labrax),* amberjack (*Seriola quinqueradiata*), turbot (*Scophthalmus maximus*), common dab (*Limanda limanda*), Atlantic halibut (*Hippoglossus hippoglossus)*, common sole (*Solea solea*), Senegalese sole (*Solea senegalensis*) and Altantic cod (*Gadus morhua)*.

[Susceptible Stage] Juveniles less than 6 months old are most severely affected. At water temperature of 10~12℃ , the morbidity and mortality rates can reach 20%~100% in rainbow trout fry at 2~10 weeks of age. Clinical disease normally does not occur in fingerling after 20 weeks of age, but they can become life-long carriers.

[Outbreak Water Temperature] Disease outbreaks generally occur at 10~15℃. Adult salmonids and non-salmonids such as carp *(Cyprinus carpio)* and amur pike (*Esox reicherti*), etc., can become sub-clinical carriers.

[Geographic Distribution] Endemic in Europe, Asia, North America and South Africa.

Aetiological agent

(1) Infectious pancreatic necrosis virus, IPNV.

(2) Family: *Binaviridae*. Genus: *Aquabirnavirus*.

(3) The virion is non-capsulated, 50~75nm in diameter, symmetrical icosahedral with a capsid.

(4) The viral genome is composed of two linear double stranded RNA molecules of 6kb in size, segment A and B. Segment A consists of two open reading frames (ORF), the smaller one encodes for a non-structural protein, VP5, with a molecular weight of 16,500u; the larger one encodes for a polymeric protein, which can form VP2, VP3 and VP4. Segment B is 2.8kb in size encoding RNA polymerase VP1 with a molecular weight of 94ku, existing as a genome binding protein (VPg) connecting the ends of Segments A and B.

(5) 10 serotypes are identified, including group A serotypes: A1 (West Buxton, WB), A2 (Spajarup, Sp), A3 (Abild, Ab) A4 (Hecht, He), A5 (Tellina, Te), A6 (Canada3, C3) and A9 (Jasper, Ja), and group B serotype: B1 (Tellinavirus, TV-1). Serotype A1 is mainly present in the United States of America, serotypes A6~A9 are mainly present in Canada, while serotypes A2~A5 and B1 are mostly present in Europe and Asia.

Clinical signs and pathological changes

(1) Affected fish produce mucoid excretions and lost co-ordination in swimming, they often swim in vertical spiral motion until death.

(2) Darkened body color, exophthalmia, abdominal distension and ascites.

(3) Inappetence, empty gastrointestinal tract upon post-mortem examination, hard yellowish

materials or greyish-white catarrhal discharge may be present in the guts. Petechiae in pyloric stomach, caeca and cranial body fat, reproductive organs and other viscera.

(4) Histopathological findings include diffuse necrosis in the pancreas and also in the kidney.

Diagnostic methods

(1) **Virus isolation** Use BF-2, EPC, RTG-2, FHM and CHSE-214 cell lines, and incubate at 15℃.

(2) **Reversed transcriptase-polymerase chain reaction (RT-PCR)** The primers are IPNV-F (5′-CCA-GCG-AAT-ATT-TTC-TCC-ACC-A-3′) and IPNV-R (5′-AGG-AGA-TGA-CAT-GTG-CTA-CAC-CG-3′) with annealing temperature at 55℃. The amplicon size is 224bp. Perform sequencing on the amplicon for confirmation.

(3) **Neutralisation test** Perform serial dilution on the virus culture media and respectively add IPNV reference antiserum, and cell culture fluid and incubate at 20℃. Calculate the neutralization index. If the index is higher than 50, it could be suspected as IPNV infection.

(4) **Enzyme-linked immunosorbant assay (ELISA)**

Preventative measures

(1) Maintain good hygiene in aquaculture farms.

(2) Apply stringent quarantine measures on broodstock, fry and fingerling introduced to the farm.

(3) Select SPF fry or fingerling for farming.

(4) Fish eggs can be disinfected with 50g/m^3 Iodophor PVP-1 for 15 minutes

(5) At early course of disease, PVP-I can be applied in feed at 1.64~1.91g of effective iodine per kg of fish daily for 15 consecutive days.

(6) Oral administration of immunostimulant and metabolic modifiers, such as polysaccharides, polypeptide, multivitamins, etc.

(7) Use commercial vaccine.

患IPN病鱼临床症状及病变

A. 头朝上、尾朝下垂直悬浮于水体中

B. 患病斑点叉尾鮰腹部膨大，眼球突出

[源自冯刚及汪开毓]

Clinical signs and macroscopic findings of affected channel catfish

A. Affected fish are showing upside down swimming

B. Marked abdominal distension and exophthalmia of affected fish

[Source: Gang Feng and Kaiyu Wang]

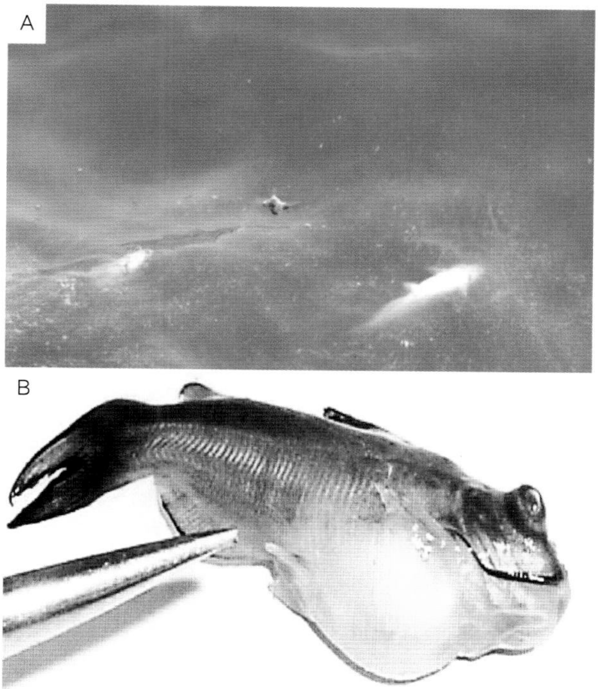

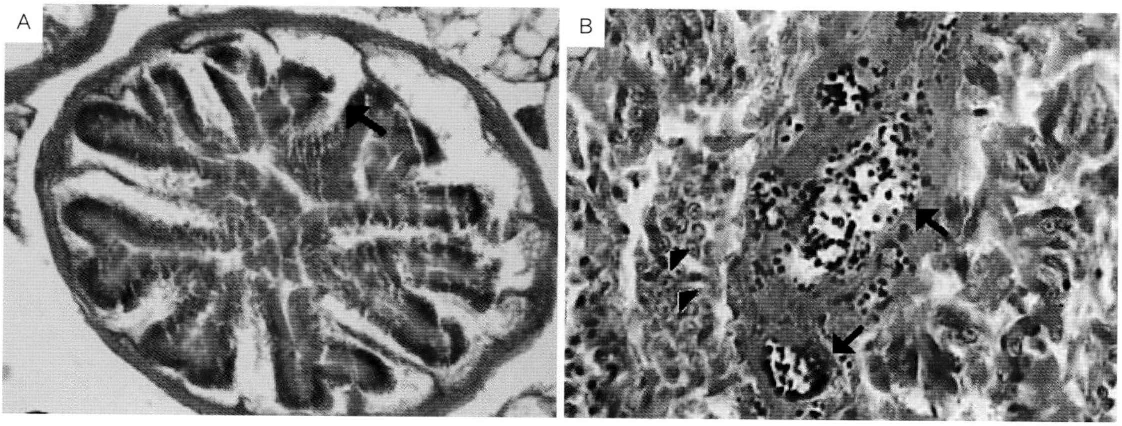

患IPN病鱼的细胞及组织病理病灶（HE染色）

A. 胰脏急性坏死（箭头） B. 肠道细胞坏死（箭头）及核固缩及崩解（三角形）

[源自Gabriel Aguirre-Guzmán, Ned Iván de la Cruz-Hernández and Jesús Genaro Sánchez-Martínez]

Histological lesions of affected fish (HE staining)

A. Acute necrosis in pancreas (arrow) B. Generalized necrosis in intestine (arrow) and cells with nuclear pyknosis and karyorrhexis (arrow head)

[Source: Gabriel Aguirre-Guzmán, Ned Iván de la Cruz-Hernández and Jesús Genaro Sánchez-Martínez]

斑点叉尾鮰病毒病

疾病概述

【概述】 斑点叉尾鮰病毒病是一种严重的急性致死性传染病。

【宿主】 自然暴发于斑点叉尾鮰（*Letalurus punetaus*）的鱼苗和鱼种，也可人工注射感染白叉尾鮰、长鳍叉尾鮰（*Lctalurus furcatus*）、斑点叉尾鮰与长鳍叉尾鮰杂交种。此外，CCV还感染南方大口鲇（*Silurus meridionalis*）和大口黑鲈（*Micropterus salmoides*），被感染的鱼迅速死亡。

【易感阶段】 此病多发于1龄鱼种，成鱼也可发生隐形感染成为病毒携带者。

【发病水温】 常发生于水温在25℃以上的水域，且偶发性高。在水温25 ～ 30℃时广泛流行，传播速度快，病程短，死亡率高，在一些养殖场甚至有100%的死亡率。当温度降低至18℃时，死亡率明显降低，流行程度与温度成正比。

【地域分布】 20世纪60年代，最先在美国流行且陆续在其他地方暴发，主要流行于北美地区，东南亚及西欧部分国家和地区也有分布。

病原

（1）病原是斑点叉尾鮰病毒（Channel catfish virus，CCV），又称鮰疱疹病毒Ⅰ型（*Ictalurid herpesvirus*-Ⅰ）。

（2）属异疱疹病毒科（*Alloherpesviridae*）、鮰鱼疱疹病毒属（*Ictalurivirus*）。

（3）病毒粒子呈二十面体，直径为175 ～ 200nm，有囊膜。

（4）为双股DNA，基因组全长为134.2 kb，含有18.5kb的两端正向2重复序列和1个97.1kb的特异性序列（UL）。整个基因有77个开放阅读框（ORFs），其中14个位于两端的重复序列中，共编码32个多肽，其中，有18个结构蛋白。病毒分子质量约85×10^6ku，病毒多肽分子质量为12 000 ～ 300 000ku。

临床症状和病理学变化

（1）摄食活动减弱，离群独游，反应迟钝，不规则游动、打转，其间或有短暂的激烈活动，后期，大量病鱼聚集在孵化池和池塘边缘，头朝上、尾朝下悬挂着，并沉入水底而死亡。

（2）眼部突出，表皮发黑，鳃发白，鳍条和肌肉出血，腹部膨大，肛门红肿外突。

（3）肾脏、肝脏、脾脏、分泌组织、胃肠道、胰腺和骨骼肌出现聚集或扩散坏死、出血或有瘀斑。神经细胞核出现空泡，周围神经纤维肿大。

诊断方法

（1）病毒分离 使用斑点叉尾鮰卵巢细胞系（CCO）或棕鮰细胞系（BB），分离温度为25 ～ 30℃。

（2）PCR 引物为：

① CCV-F（5′-GGC-TTG-GGC-TTG-ATG-GAC-3′）和CCV-R（5′-GAG-GAG-GAC-AAC-GCG-ACT-G-3′），退火温度为60℃，扩增产物长度为144bp。

② CCV 8F（5′-ACG-TGT-ATC-ACG-GTC-TCA-CT-3′）和CCV 8R（5′-TTC-GAG-AAT-CGG-GTC-TCT-GT-3′），退火温度为55℃，扩增产物长度为335 bp。

③ CCV 59F（5′-GTT-CTG-AAG-AAC-GCC-CTG-AA-3′）和CCV 59R（5′-AAC-ACC-AGT-ATC-ACG-AGG-AG-3′），退火温度为55℃，扩增产物长度为329bp。

以上扩增产物经测序后进行判定。

（3）酶联免疫吸附试验（ELISA）

防治方法

（1）保持养殖场良好的卫生管理水平。

（2）对养殖场引入的亲鱼或苗种，采取严格的检疫和隔离措施。

（3）对池塘进行消毒，对网具、运输工具等用高浓度消毒液（5%的甲醛或40mg/L的有效氯）浸泡消毒，防止病毒传入或扩散。

（4）鱼苗放养密度不宜过大，最好每亩低于8 000尾。对于孵化池，保持最佳水质，尤其要注意保持较高的溶解氧水平，池水溶解氧应保持在4 mg/L以上，有利于减少鱼的应激反应。同时，保持水体的良好循环，及时清除水体中的杂物。

（5）对于育苗池，要保持合理的养殖密度，合理投饲。

（6）在投鱼投饲等操作鱼苗和鱼种时，最好在水温20℃以下进行；在鱼病暴发时，尽量不要进行捕鱼等作业，减少鱼的应激。

（7）用0.1mg/L的强氯精，全池泼洒，连用2d。

（8）每100kg饲料加0.1kg的三黄粉（黄芩、黄连、黄柏）制成药饵，连喂5 ～ 7d。

Channel catfish virus disease, CCVD

Disease overview

[Disease Characteristic] Severe, acute and lethal contagious disease.

[Susceptible Host] The disease naturally occur in fry and finerglings of channel catfish (*Letalurus punetaus*), while blue catfish (*Lctalurus furcatus*) and hybrid of blue catfish and channel catfish can also be infected by artificial injection. In addition, CCV can also infect Chinese large-mouth catfish (*Silurus meridionalis*), and large mouth bass (*Micropterus salmoides*) causing rapid death.

[Susceptible Stage] Mostly occur in juveniles at 1 year of age, while adult fish can be subclinical carriers.

[Outbreak Water Temperature] Disease outbreaks occur sporadically when the water temperature is above 25 ℃. When water temperature is between 25~30 ℃, the disease is endemic and propagates rapidly with short clinical disease courses and high mortality. The mortality rate can reach 100% in some fish farms. When the temperature drops below 18℃ , the mortality decreases significantly. The prevalence of the disease is directly proportional to the water temperature.

[Geographic Distribution] CCVD was first found to be endemic in America in 1960's and then outbreaks started in other places, it is mainly endemic in most of the countries in the North America, South East Asia and some countries in Eastern Europe.

Aetiological agent

(1) *Ictalurid herpesvirus*-Ⅰ, also called Channel catfish virus (CCV).

(2) Family: Alloherpesviridae. Genus: *Ictalurivirus*.

(3) The virion is enveloped, icosahedron in shape, and 175~200nm in diameter.

(4) The viral genome is composed of a double stranded DNA. The whole genome is 134.2kb in size, containing 18.5kb of two forward repeats and a 97.1kb specific sequence (UL). There is a total of 77 Open Reading Frames (ORFs) in the viral genome. Among these ORFs, 14 of them located at the repeats at both ends, encoding 32 polypeptides, 18 of which are structural proteins. The molecular size of the virus is around 85×10^6 ku, and the molecular weight of the viral polypeptides is 12,000~300,000ku.

Clinical signs and pathological changes

(1) Reduced appetite, segregated, slow and irregular movement, spiraling with occasional fierce behaviors. At later stage, large number of affected fish will gather at the edge of the hatcheries or pond in a head-up and tail-down "hanging" position, sinking to the bottom before death.

(2) Exophthalmia, darkening of body color, pale gill, haemorrhage at fins and muscles, abdominal distension, anal swelling and protrusion.

(3) Focal or diffuse necrosis, haemorrhage or bruising in kidney, liver, spleen, secretory tissues, intestinal tract, pancreas and skeletal muscles. Vacuolation of the nuclei in the nervous tissues with swollen nerve fibers in the surrounding.

Diagnostic methods

(1) **Virus isolation** Using CCO or BB cell-lines and incubate at 25~30℃ .

(2) **Polymerase chain reaction (PCR)** There are 3 pairs of primers available.

① CCV-F (5′-GGC-TTG-GGC-TTG-ATG-GAC-3′) and CCV-R (5′-GAG-GAG-GAC-AAC-GCG-ACT-G-3′) with annealing temperature at 60℃ . The amplicon size is 144bp.

② CCV 8F (5′-ACG-TGT-ATC-ACG-GTC-TCA-CT-3′) and CCV 8R (5′-TTC-GAG-AAT-CGG-GTC-TCT-GT-3′) with annealing temperature at 55℃ . The amplicon size is 335bp.

③ CCV 59F (5′-GTT-CTG-AAG-AAC-GCC-CTG-AA-3′) and CCV 59R (5′-AAC-ACC-AGT-ATC-ACG-AGG-AG-3′) with annealing temperature at 55℃ . The amplicon size is 329bp.

Perform sequencing on the amplicons for confirmation.

(3) **Enzyme-linked immunosorbent assay (ELISA)**

Preventative measures

(1) Maintain good hygiene management in the fish farm.

(2) Apply stringent quarantine measures on fry and broodstock introduced to the farm.

(3) Disinfect the pond and equipment such as nets and transport tools by soaking with high concentration disinfectants (5% formalin or 40mg/L available chlorine) to prevent introduction or spread of virus.

(4) Maintain appropriate cultivation density of fry, best to be within 8,000 fries per 667m^2. Maintain good water quality in hatcheries especially the dissolved oxygen level (above 4mg/L) to reduce stress of fish. Maintain good water circulation and remove contaminants timely.

(5) Maintain appropriate stocking and culturing density of the farm and provide appropriate amount of feed.

(6) For stocking and feeding, best to carry out when the water temperature is under 20 ℃ . When there is an outbreak, avoid operations such as capture to reduce any potential stress to the fish population.

(7) Trichloroisocyanuric acid: Apply trichloroisocyanuric acid to the whole pond at 0.1mg/L for 2 consecutive days.

(8) Use medicated feed by mixing 0.1kg of "San Huang" powder (Chinese skullcap, Chinese goldthread, and Cortex Phellodendri) with 100kg of feed to treat for 5~7 consecutive days.

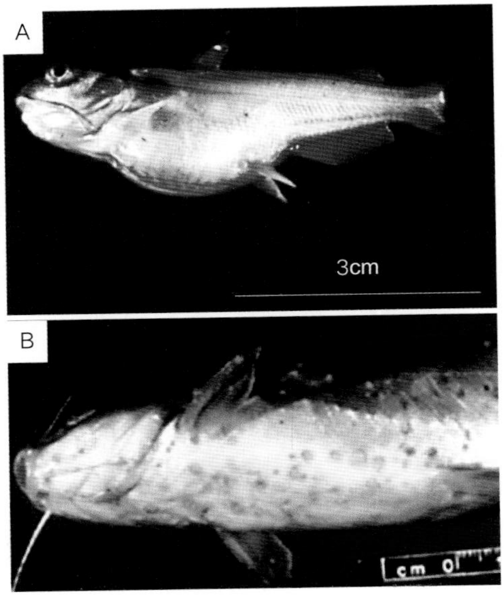

感染CCVD的鮰

A．幼鱼：腹部膨大，眼睛突出

B．成鱼：体表、鳃、鳍上有出血点

[源自L．A．汉森及美国农业部]

Macroscopic findings of affected fish

A．Affected catfish fingerling. Exophthalmia and abdominal distension

B．Affect adult catfish. Petechiae and ecchymosis on the body surface, gill and fins

[Source：L．A．Hanson and United States Department of Agriculture]

弧菌病

疾病概述

【概述】 弧菌病是海水鱼类最常发生的细菌性疾病。

【宿主】 海水鱼类、贝类以及甲壳类，如日本鳗鲡（*Anguilla japonica*）、对虾（*Penaeus ori-entalis*）、尖吻鲈（*Lates calcarifer*）、真鲷（*Pagrosomus major*）、大西洋鲑（*Salmo salar*）、竹筴鱼（*Trachurus japonicus*）、比目鱼（Pleuronectiformes）等。

【易感阶段】 从孵化后数月到1龄左右的鱼易感染，多为慢性。

【发病水温】 水温15 ～ 25℃时是发病高峰期。

【地域分布】 弧菌耐低温和碱，外界环境中生存能力强，在水中存活长，各处井水、海水、河水中可以存活1 ～ 3周。因此，弧菌病流行于世界各地的海水养殖鱼、虾、贝类。

病原

（1）病原主要是鳗弧菌（*Vibrio anguillarum*）、副溶血弧菌（*V. parahaemolyticus*），此外，还有溶藻胶弧菌（*V. alginolyticus*）、哈维氏弧菌（*V. harveyi*）、创伤弧菌（*V. vulnificus*）、杀鲑弧菌（*V. salmonicida*）、海利斯顿氏菌（*V. pelagius*）、美人鱼弧菌（*V. damsela*）、奥氏弧菌（*V. ordalii*）、费氏弧菌（*V. fischeri*）、鲨鱼弧菌（*V. carchariae*）以及拟态弧菌（*V. mimicus*）等。通过细菌分类学已经证实鲨鱼弧菌和哈维氏弧菌是同一种菌。

（2）属弧菌科（Vibrionaceae）、弧菌属（*Vibrio*）。

（3）呈短杆状，稍弯曲，两端圆形，大小为（0.5 ～ 0.7）μm ×（1 ～ 2）μm，无荚膜，有一端生1根鞭毛或更多根鞭毛。革兰氏阴性，需氧或兼性厌氧菌，运动活泼。

（4）鳗弧菌有5种生物型，即鳗弧菌生物A、B、C、D、E型，能引起世界范围内50多种淡、海水养殖鱼类及其他养殖动物发生弧菌病。同时，鳗弧菌有23种血清型，O1、O2及部分O3血清型为主要的致病血清型，O2a和O2b血清亚型是引起大西洋鳕鱼弧菌病的主要病原，其他血清型为环境菌株，对鱼类没有致病性。创伤弧菌有3种生物型，分别是生物Ⅰ型、生物Ⅱ型和生物Ⅲ型。生物Ⅱ型是海水养殖经济动物的致病菌，主要对日本鳗鲡、对虾致病，部分生物Ⅲ型菌株可对淡水养殖的罗非鱼致病。

临床症状和病理学变化

（1）食欲不振，缓慢地浮游于水面，有时回旋状游泳。

（2）不同病原、不同寄主，症状不同，共同病症是体表皮肤溃疡。感染初期，体色多呈斑块状褪色；中度感染，鳍基部、躯干部等发红或出现斑点状出血。

（3）鳞片脱落，吻端、鳍膜烂掉，眼内出血，肛门红肿扩张，常有黄色黏液流出。

（4）病灶组织浸润呈出血性溃疡。

诊断方法

（1）细菌分离 使用2216E和TCBS培养基，适宜温度培养24 ~ 48h，两种弧菌的菌落在2216E培养基上呈圆形、稍凸、边缘平滑、灰白色、略透明、有光泽。其中副溶血弧菌在TCBS培养基上可形成2mm以上的蓝绿色菌落，而鳗弧菌在TCBS培养基上可形成1mm以上的黄色菌落。

（2）理化生化鉴定

（3）PCR 引物为：

① 16S rRNA基因PCR扩增引物为：27F（5′-AGA-GTT-TGA-TC（C/A）-TGG-CTC-AG-3′）和1 492R（5′-GGT-TAC-CTT-GTT-ACG-ACT-T-3′），退火温度为55℃，扩增产物长度为1 445bp。

② 16S-23S *ARNr*基因PCR扩增引物分别是16S F（5′-CTC-CCC-TAC-GGG-ATA-CCA-TT-3′）和16S R（5′-TTG-TAA-GGC-AGG-TGC-TCT-CC-3′），退火温度为60℃，扩增产物长度为580bp。

③ 16S rDNA基因PCR扩增的引物为：F（5′-CGG-TGA-AAT-GCG-TAG-AGA-T-3′）和R（5′-TTA-CTA-GCG-ATT-CCG-AGT-TC-3′），退火温度为63℃，扩增产物长度为750bp。

以上扩增产物经测序后判定。

（4）Real-time PCR

①副溶血弧菌的pR72 DNA基因扩增的引物分别是pR72 F（5′-CGC-TTT-AAA-ACA-CCG-TCA-GC-3′）和pR72 R（5′-GAA-GAT-TAC-CCG-CTT-GCT-GT-3′），探针pR72 P（5′-Hex-GCG-TGT-GAT-TTT-TCT-CAC-GA-BHQ-3′），退火温度为57.8℃，扩增产物长度为74bp。

②溶藻弧菌的gyrase B（gyrB）基因扩增的引物分别是gyrB F（5′-CGC-CTA-AAG-CTC-GTG-AAA-TG-3′）和gyrB R（5′-GAG-TGC-CGG-ATC-TTT-TTC-CT-3′），探针为gyrB P（5′-ROX-TAG-ACC-TAG-CAG-GCC-TTC-CA-BHQ2-3′），退火温度为57.8℃，扩增产物长度为98bp。

③霍乱弧菌的toxR扩增的引物分别是toxR F（5′-TCA-AGC-AGT-GTG-CCT-TCA-TC-3′）和toxR R（5′-CCA-AGT-TTG-GAG-CCG-ATT-TA-3′），探针为toxR P（5′-FAM-TGT-AGT-GAA-CAC-ACC-GCA-GC- BHQ 1-3′），退火温度为57.8℃，扩增产物长度为85bp。

防治方法

（1）保持优良的水质和养殖环境，不投喂腐败变质的小杂鱼、虾。

（2）投喂磺胺类药物饵料——磺胺甲基嘧啶，第一天每千克鱼体重用药200mg，第二天以后减半，连续投喂7 ~ 10d（注意：磺胺类药物在鳗饲料添加剂中已被禁止使用）。

（3）投喂抗生素药饵，如土霉素，每千克鱼体重每天用药70 ~ 80mg，制成药饵，连续投喂5 ~ 7d。

（4）在口服药饵的同时，用漂白粉等消毒剂全池泼洒，视病情用1 ~ 2次，可以提高防治效果。

（5）使用商品化鳗弧菌疫苗，采用注射、口服、浸泡和喷雾等方法进行免疫。

Vibriosis

Disease overview

[Disease Characteristic] Vibriosis is the most common bacterial disease in marine fish.

[Susceptible Host] Mainly affect marine fish, molluscs and crustaceans, such as Japanese eel (*Anguilla japonica*), oriental shrimp (*Penaeus orientalis*), barramundi (*Lates calcarifer*), red seabream (*Pagrosomus major*), Atlantic salmon (*Salmo salar*), Japanese horse mackerel (*Trachurus japonicus*), flatfish (Pleuronectiformes), etc.

[Susceptible Stage] Fish are more susceptible from a few months after hatching to about one year of age. Disease is mostly chronic.

[Outbreak Water Temperature] Peak disease outbreaks occur when water temperature is 15~25℃ .

[Geographical Distribution] *Vibrio* spp. are tolerant to low temperature and alkaline, they survive well in the environment for a long period in water up to 1~3 weeks in well water, seawater or river water. Therefore, vibriosis is endemic globally in aquaculture of marine fish, shrimp and molluscs.

Aetiological agents

(1) The major aetiological agents are *V. anguillarum and V. parahaemolyticus*. Other species include *V. alginolyticus*, *V. harveyi*, *V. vulnificus*, *V. salmonicida*, *V. pelagius*, *V. damsela*, *V. ordalii*, *V. fischeri*, *V. carchariae* and *V. mimicus*, etc. Bacterial taxonomical analysis has been confirmed that *V. carchariae* and *V. harveyi* are the same bacteria species.

(2) Family: Vibrionaceae. Genus: *Vibrio*.

(3) Short curved-rod, obtuse at both ends, (0.5~0.7)μm ×(1~2)μm in size, non-capsulated, with one or more polar flagella at one end. Gram-negative, aerobic or facultative anaerobic, highly motile.

(4) *V. anguillarum* has 5 biotypes–biotype A, B, C, D and E. It can cause vibriosis globally in over 50 species of fish cultured in fresh water and marine water, and other aquaculture animals. Concurrently, *V. anguillarum* is consist of 23 serotypes, O1, O2 and some O3 serotypes strains and pathogenic, and sub-serotype of O2a and O2b are the main cause of vibriosis in Alantic cod (*Gadus morhua*). Other serotypes are environmental strains and have no pathogenicity to fish.*V. vulnificus* has 3 biotypes, namely biotypes Ⅰ, Ⅱ and Ⅲ. Biotype Ⅱ is a pathogen for seawater aquaculture, mainly affecting Japanese eel and oriental shrimp. Some biotype Ⅲ strains can affect tilapia cultured in freshwater.

Clinical signs and pathological changes

(1) Inappetence, slow movement near the surface of the water, sometimes whirling.

(2) Different pathogens and different hosts may show different clinical signs, a common one is skin ulcers. In the early stage of infection, patchy fading of the body color is common. In moderate infection, ecchymosis or petechiae may be present at the base of the fin, body trunk, etc.

(3) Scales loss, snout and fin necrosis, intraocular haemorrhage, anal swelling and reddening, often with yellow mucoid discharge.

(4) Infiltration of affected tissues and appearing as hemorrhagic ulcerations.

Diagnostic methods

(1) **Bacterial isolation** Use 2216E and Thiosulfate Citrate Bile Salts Sucrose (TCBS) media and incubate for 24~48 hours at appropriate temperature. The colonies of *Vibrio* spp. in 2216E medium are round, slightly convex, smooth edged, grayish white in color, slightly translucent and glossy. *V. parahaemolyticus* can form blue-green colonies over 2mm in TCBS medium, while V. anguillarum can form yellow colonies over 1mm in TCBS medium.

(2) **Biochemical identification**

(3) **Polymerase chain reaction (PCR)** Three pairs of primers are available.

① 16S rRNA PCR primers: 27F (5′-AGA-GTT-TGA-TC(C/A)-TGG-CTC-AG-3′) and 1,492R (5′-GGT-TAC-CTT-GTT-ACG-ACT-T-3′) with annealing temperature at 55℃. The amplicon size is 1,445bp.

② 16S-23S *ARNr* PCR primers: 16S F (5′-CTC-CCC-TAC-GGG-ATA-CCA-TT-3′) and 16S R (5′-TTG-TAA-GGC-AGG-TGC-TCT-CC-3′), with annealing temperature at 60℃. The amplicon size is 580bp.

③ 16S rDNA primers: forward primer (5′-CGG-TGA-AAT-GCG-TAG-AGA-T-3′) and reverse primer (5′-TTA-CTA-GCG-ATT-CCG-AGT-TC-3′), with annealing temperature at 63℃. The amplicon size is 750bp.

Perform sequencing on the amplicons for confirmation.

(4) **Real-time PCR**

① *V. parahaemolyticus* pR72 DNA primers: pR72 F (5′-CGC-TTT-AAA-ACA-CCG-TCA-GC-3′) and pR72 R (5′-GAA-GAT-TAC-CCG-CTT-GCT-GT-3′) with probe pR72 P (5′-Hex-GCG-TGT-GAT-TTT-TCT-CAC-GA-BHQ-3′), The annealing temperature is 57.8℃ and the amplicon size is 74bp.

② *V. alginolyticus* gyrase B (gyrB) primers: gyrB F (5′-CGC-CTA-AAG-CTC-GTG-AAA-TG-3′) and gyrB R (5′-GAG-TGC-CGG-ATC-TTT-TTC-CT-3′) with probe gyrB P (5′-ROX-TAG-ACC-TAG-CAG-GCC-TTC-CA-BHQ2-3′), The annealing temperature is 57.8℃ and the amplicon size is 98bp.

③ *V. cholera* toxR primers: toxR F (5′-TCA-AGC-AGT-GTG-CCT-TCA-TC-3′) and toxR R (5′-CCA-AGT-TTG-GAG-CCG-ATT-TA-3′) with probe toxR P (5′-FAM-TGT-AGT-GAA-CAC-ACC-GCA-GC-BHQ1-3′), The annealing temperature is 57.8℃ and the amplicon size is 85bp.

Preventative measures

(1) Maintain good water quality and aquaculture environment; do not use deteriorated fish baits and prawns for feeding.

(2) Sulfonamides feed treatment: sulfamerazine at 200 mg/kg of fish on the first day, half dose afterwards, mix in feed to treat for consecutive 7~10 days. (Note: sulfonamides are banned for use as eel feed additives.)

(3) Oxytetracycline feed treatment: 70~80mg/kg of fish daily, mix in feed to treat for consecutive 5~7 days.

(4) While on treatment with medicated feed, disinfecting the pond with disinfectant, such as bleaching powder once or twice depending on actual situation, can improve treatment effectiveness.

(5) Use commercial *V. anguillarum* vaccine, immunize the fish via injection, oral, immersion or spraying routes.

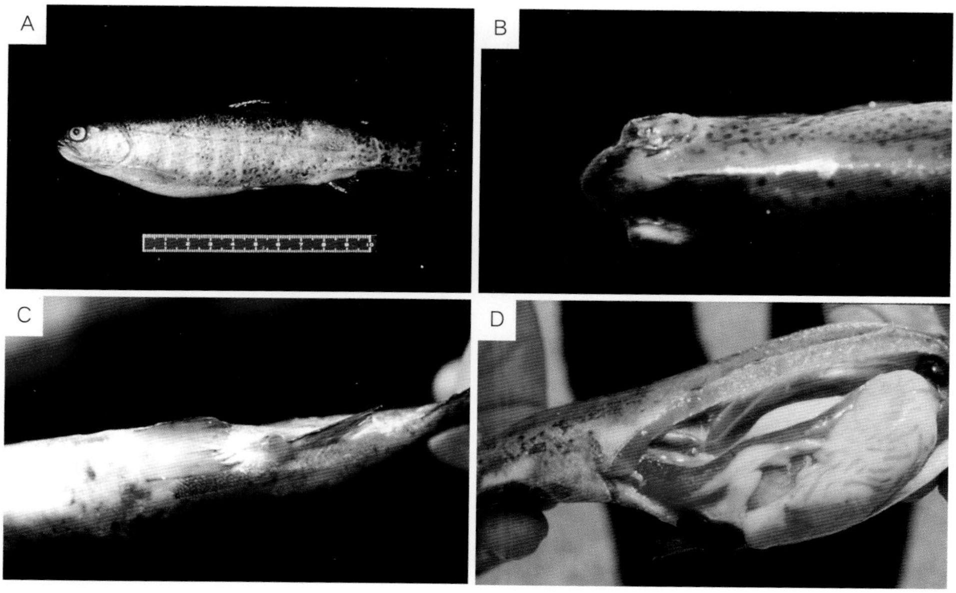

患鱼病变

A. 体色变黑　B. 眼球突出、发红　C. 鳍条基部、体侧部发红，肛门发红、扩张

D. 肠管卡他性炎症

[源自《新鱼病图鉴》，小川和夫]

Macroscopic findings of affected fish

A. Darkened body color　B. Exophthalmia and reddening of periocular tissues　C. Ecchymosis of fin base and fin rays. Reddening and swelling of the anus　D. Catarrhal inflammation of the intestines

[Source: *New Atlas of Fish Diseases*, Kazuo Ogawa]

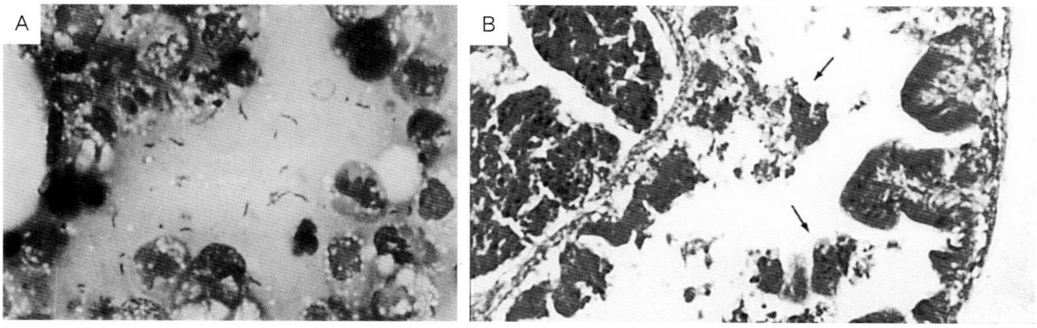

病鱼细胞及组织病变

A．刘氏染色高倍镜下，可见弧形菌体（刘氏染色，1 000倍）

B．感染鱼肠道弧菌(*Vibrio ichthyoenteri*)的病鱼肠管黏膜上皮细胞的排列紊乱、脱落（箭头）(HE染色）

[源自《新鱼病图鉴》，小川和夫]

Cytology findings and histological lesions of affected fish

A short-curved *Vibrio* spp. present in impression smear of the lesion (Liu's staining, 1,000 ×)

B．Infection of *Vibrio ichthyoenteri* causing sloughing and distortion of the normal architecture of intestinal epithelial cells (arrow) (HE staining)

[Source：*New Atlas of Fish Diseases*，Kazuo Ogawa]

疖疮病

疾病概述

【概述】 疖疮病是一种主要感染鲑鳟的细菌性流行病。

【宿主】 主要感染虹鳟（*Oncorhynchus mykiss*）、大西洋鲑（*Salmo salar*）、白斑狗鱼（*Esox lucius*）、大菱鲆（*Scophthalmus maximus*）、半滑舌鳎（*Cynoglossus semilaevis*），也感染鲤（*Cyprinus carpio*）、鲢（*Hypophthalmichthys molitrix*）、鲫（*Carasstus anratus*）、斑点叉尾鮰（*Ietalurus punetaus*）、刺参（*Apostichopus japonicas*）、乌鳢（*Channa argus*）、海七鳃鳗（*Lampetra japonica*）、细鳞鲑（*Brachymystax lenok*）等。

【易感阶段】 一般感染高龄鱼，鱼苗、夏花未见有患此病。

【发病水温】 流行无季节性，一年四季均可发生，一般为散发性。

【地域分布】 主要发生于欧洲、北美洲、日本、中国等地。

病原

（1）病原为杀鲑气单胞菌（*Aeromonas salmonicida*）。

（2）属气单胞菌科（Aeromonadaceae）、气单胞属（*Aeromonas*）。

（3）两端圆形，大小为$(0.8 \sim 2.1)\mu m \times (0.35 \sim 1.0)\mu m$。单个或两个相连，极端单鞭毛，有荚膜，无芽孢，革兰氏阴性短杆菌，需氧或兼性厌氧。

（4）在肠道选择培养基上，大多数菌株形成乳糖不发酵菌落；在TCBS琼脂上生长不良。在液体培养基中呈均匀混浊。生长的适宜温度为25℃，50℃时微弱生长，65℃时1h死亡。

（5）杀鲑气单胞菌的毒力因子按照功能可分为七大类，包括外毒素、胞外酶、黏附因子、分泌系统、Fe摄取系统、抗药基因和群体感应系统。其中，气溶素、溶血素、蛋白酶、表面阵列蛋白、菌毛、Ⅲ型分泌系统、Fe载体和群体感应效应分子受体等为杀鲑气单胞菌主要毒力因子。

临床症状和病理学变化

（1）病发鱼体分三型：急性型病鱼急性死亡，无外部症状；亚急性型病鱼病情发展较慢，在躯干肌肉形成疖疮，有外部症状，陆续死亡；慢性型病鱼长期处于带菌状态，无症状也不死亡。

（2）病鱼体色发黑，食欲减退，鱼体消瘦，游动缓慢，常离群独游。

（3）患病鱼背鳍基部的皮肤及肌肉组织发炎，红肿处凸出体表，触之有流动感，患部有灰白脓液流出。生出1个或几个与人类疖疮病相似的脓疮。在皮下肌肉内形成病灶，随着病灶内细菌繁殖增多，皮肤肌肉发炎，化脓形成脓疮，脓疮内部充满脓汁、血液和大量细菌。

（4）患部软化，向外隆起，隆起的皮肤先是充血，然后出血，继而坏死、溃烂，形成火山形的溃疡口。切开患处，可见肌肉溶解，呈灰黄色的混浊或凝乳状，有灰白色或红灰色脓液流出，肠道充血发炎，其他内脏器官未见肉眼可见病变。

（5）病理组织切片可见患处的真皮发生肿胀、变性、充血、出血，但尚未坏死。病灶中心的骨骼肌纤维已完全解体，在其中可见大量杆菌、脓液及少量已坏死、解体的炎症细胞。有弥漫性化脓性炎，即蜂窝织炎。

诊断方法

（1）细菌分离 使用血琼脂培养基，25℃培养24h后，形成灰白、光滑、湿润、凸起的直径约2mm的菌落，多数菌株有β溶血环，3～5d后菌落呈暗绿色。

（2）理化生化鉴定 革兰氏阴性，氧化酶阳性，过氧化氢酶阳性，发酵葡萄糖，在TSA培养基上不产生水溶性褐色色素，可产生吲哚，不产生H_2S，可分解蔗糖、半乳糖，不能分解甘露醇，V-P反应阴性。

（3）PCR 有3组引物可供选择：

① fD1（5′-AGA-GTT-TGA-TCC-TGG-CTC-AG-3′）和rp2（5′-ACG-GCT-ACC-TTG-TTA-CGA-CTT-3′），退火温度为56℃，扩增产物长度约1 500bp。

② MIY1（5′-AGC-CTC-CAC-GCG-CTC-ACA-GC-3′）和MIY2（5′-AAG-AGG-CCC-CAT-AGT-GTG-GG-3′），退火温度为60℃，扩增产物长度为512bp。

③ 27F［5′-AGA-GTT-TGA-TC（C/A）-TGG-CTC-AG-3′］和1 492R（5′-GGT-TAC-CTT-GTT-ACG-ACT-T-3′），退火温度为55℃，扩增产物长度为1445bp。

以上扩增产物经测序后判定。

防治方法

1.预防措施

（1）在捕捞、运输、放养等操作过程中，尽量避免鱼体受伤。鱼种放养前可用3%～4%浓度的食盐浸5～15min或5～8mg/L的漂白粉溶液浸洗20～30min。药浴时间的长短，视水温和鱼体忍受力而灵活掌握。

（2）用微生态制剂改良水质，可按每667m^2（水深1m）首次施用量1kg，以后每隔15d施用1次，用量0.5kg。也可用常规使用的生石灰溶液、沸石粉等水质改良剂优化水质。

（3）用硫酸铜、敌百虫等杀虫剂杀灭水中鱼体外寄生虫，以防寄生虫侵害鱼体，使鱼体受损而感染此病。

（4）鱼种按时注射疫苗，增强免疫力，注射疫苗后可用3%～5%的食盐水或20g/m^3的高锰酸钾液浸泡10～15min，以避免继发感染。

（5）如少量鱼体受外伤，可用新鲜大蒜汁涂擦2～3次。

（6）养殖期间，每隔15d按0.3～0.5g/m^2的二氧化氯全池泼洒。

（7）将中药五倍子捣碎，用开水全部溶解后，用池水稀释全池泼洒，用量为2 ～ 4g/m^2。还可每667m^2（水深1m）用大青叶5kg加黄连1kg，加水3次煎汁至药液为20kg，全池泼洒，连用3d。

2.治疗方法

（1）用0.2 ～ 0.3mg/L的二氯海因，全池泼洒，在疾病流行季节每15d 1次。

（2）每千克鱼体重每天用四环素40 ～ 80mg，或氟苯尼考5 ～ 15mg，拌饲投喂，连用3 ～ 5d。

（3）用磺胺嘧啶拌饲料投喂，第一天用量是每千克鱼体重用药100mg，以后每天用药50mg，连喂1周。方法是把磺胺嘧啶拌在适量的面糊内，然后与草料拌和，稍干一下投喂草鱼。青鱼可拌在米糠或豆饼中进行投喂。

（4）使用商品化疫苗。

Furunculosis

Disease overview

[Disease Characteristic] An endemic bacterial disease mainly affecting salmonids.

[Susceptible Host] Mainly affect rainbow trout (*Oncorhynchus mykiss*), Atlantic salmon (*Salmo salar*), northern pike (*Esox Lucius*), turbot (*Scophthalmus maximus*), Cynoglossus (*Cynoglossus semilaevis*), also affect common carp (*Cyprinus carpio*), silver carp (*Hypophthalmichthys molitrix*), goldfish (*Carasstus anratus*), channel catfish (*Ietalurus punetaus*), Japanese sea cucumber (*Apostichopus japonicas*), northern snakehead (*Channa argus*), Arctic lamprey (*Lampetra japonica*) and lenok (*Brachymystax lenok*), etc.

[Susceptible Stage] Generally affect old fish, not reported in fry and summer juvenile.

[Outbreak Water Temperature] Non-seasonal, sporadic outbreaks can occur all year round.

[Geographical Distribution] Mainly distributed in Europe, North America, Japan and China etc.

Aetiological agent

(1) *Aeromonas salmonicida*.

(2) Family: Aeromonadaceae. Genus: *Aeromonas*.

(3) Gram-negative rod-shaped aerobic or facultative anaerobic bacteria, obtuse at both ends, around (0.8~2.1) μm × (0.35~1.0) μm in size, arranged in single or in pairs, with single polar

flagella, capsulated, non-spore forming.

(4) Most strains form non-lactose fermenting colonies on intestinal bacteria selective culture media. Poor growth on Thiosulfate Citrate Bile Salts Sucrose (TCBS) agar and appears uniformly turbid in liquid media. Optimal growth at 25℃ , weak growth at 50℃ , cannot survive at 65℃ for an hour.

(5) The virulence factors of *Aeromonas salmonicida* can be divided to 7 categories according to the respective functions: exotoxins, extracellular enzymes, adhesion factors, secretion system, iron uptake system, antibiotic resistance genes and quorum sensing system. Among these, aerolysin, hemolysin, protease, surface layer protein, pilus, T3SS, iron siderophore and group effector receptor are the major virulence factors.

Clinical signs and pathological changes

(1) Infection can be divided to 3 disease types: acute disease with acute death and no external signs; sub-acute disease with slower clinical progression with external lesions and formation of furuncles in muscles on the body trunk, followed by death; and chronic disease of which the affected fish becomes persistent without any clinical sign.

(2) Darkened body color, inappetence, emaciation, slow movement, and often segregated.

(3) Inflammation of the skin and muscle tissues at the base of the dorsal fin. The inflammed areas protrude from the body surface, which are fluidity in texture, with grayish purulent discharge inside these lesions. Form one or more abscesses, resembling furunculosis in human. Initially form lesions in the subcutaneous muscles, as the bacteria propagate in these lesions, myositis and dermatitis are resulted with subsequent suppuration and abscessation. The abscesses are filled with purulent discharges, blood and large quantity of bacteria.

(4) The affected tissues soften and bulge outward with hyperaemia, haemorrhage, followed by necrosis and ulceration, forming volcano-shaped ulcers. Dissecting these areas reveals grayish yellow turbid or curd-like rhabdomyolysis with grayish white or reddish gray purulent discharge. Intestinal hyperaemia and inflammation. No observable lesions in other internal organs.

(5) Histologically, there are swelling, dysplasia, hyperaemia, and haemorrhage in the dermis, but not necrosis. The skeletal muscle fibers in the center of the lesion have completely disintegrated with numerous bacilli, purulent discharge and a few degenerated inflammatory cells, with indistinct margin between the lesion and the surrounding normal tissue. Bacteria propagate and spread in the tissues, causing infiltration of numerous inflammatory cells in the interstitium. Purulent exudate diffuses along the loose interstitium, causing diffuse suppurative inflammation, i.e. cellulitis.

Diagnostic methods

(1) Bacterial isolation Use blood agar and incubate at 25 ℃ for 24 hours. Bacterial colonies are approximately 2mm in diameter, appeared round, convex, grayish white, shinny, smooth and translucent. Most strains are beta-hemolytic, appearing dark green in color after 3 to 5 days of incubation.

(2) **Physiological and biochemical identification** Gram negative, oxidase positive, catalase positive, glucose fermentation, no production of water-soluble brown pigment on TSA medium, no H_2S production,capable to decompose sucrose and galactose but not mannitol, and V-P reaction negative.

(3) **Polymerase chain reaction (PCR)** There are 3 pairs of primers available.

① fD1 (5′-AGA-GTT-TGA-TCC-TGG-CTC-AG-3′) and rp2 (5′-ACG-GCT-ACC-TTG-TTA-CGA-CTT-3′) with annealing temperature at 56℃ . The amplicon size is 1,500bp.

② MIY1 (5′-AGC-CTC-CAC-GCG-CTC-ACA-GC-3′) and MIY2 (5′-AAG-AGG-CCC-CAT-AGT-GTG-GG-3′) with annealing temperature at 60℃ . The amplicon size is 512bp.

③ 27F [5′-AGA-GTT-TGA-TC(C/A)-TGG-CTC-AG-3′] and 1,492R (5′-GGT-TAC-CTT-GTT-ACG-ACT-T-3′) with annealing temperature at 55℃. The amplicon size is 1,445bp.

Perform sequencing on the amplicons for confirmation.

Preventative measures

(1) During capture, transport and stocking, avoid traumatizing the fish. Before stocking, the fingerlings can be treated with 3%~4% sodium chloride for 5~15min or 5~8mg/L bleaching solution for 20~30min. The duration of medicated bath depends on the water temperature and the tolerance of the fish, which needs to be monitored carefully.

(2) Use micro-ecological agents to improve water quality: First apply 1kg, followed by 0.5kg, per 667m^2 (1m depth) once every 15 days. Other common agents such as quicklime solution and zeolite powder can also be used for water quality improvement.

(3) Insecticides such as copper sulphate and trichlorfon can be used to kill ectoparasites in water to prevent damage to the fish body and subsequent bacterial infection of the wounds.

(4) Vaccination regimen to strengthen the immunity. After vaccination, immerse the fish in 3%~5% sodium chloride or 20g/m^3 potassium permanganate solution for 10~15min, to avoid secondary infection.

(5) If only a few fish are injured, use fresh garlics to scrub the wounds for 2~3 times.

(6) During culture period, apply 0.3~0.5g/m^2 chlorine dioxide to the whole pond at a 15-day interval.

(7) Dissolved grinded *Rhus chinensis* in hot water and apply to the whole pond at a concentration of 2~4g/m^2. Or prepared medicated solution per 667m^2 (1m depth) by boiling 5kg of *Folium isatidis* and 1kg of *Coptis chinesis* in hot water to come up with medicated solution of 20kg per 667m^2 (1m depth), and apply to the whole pond for 3 consecutive days.

Treatments

(1) Agribrom at 0.2~0.3mg/L, apply to the whole pond, once every 15 days in the epidemic

seasons.

(2) Tetracycline at 40~80mg/kg of fish per day, mix in feed to treat for consecutive 3~5 days. or Florfenicol at 5~15mg/kg of fish per day, mix in feed to treat for consecutive 3~5 days.

(3) Sulfadiazine. Mix with feed to treat fish at 100mg/kg of fish at the first day, followed by 50mg/kg of fish per day consecutively for 1 week. Mix sulfadiazine with appropriate amount of batter and forage followed by drying to feed grass carp. Black carp can be fed by mixing with bran or bean cake.

(4) Use commercial vaccine.

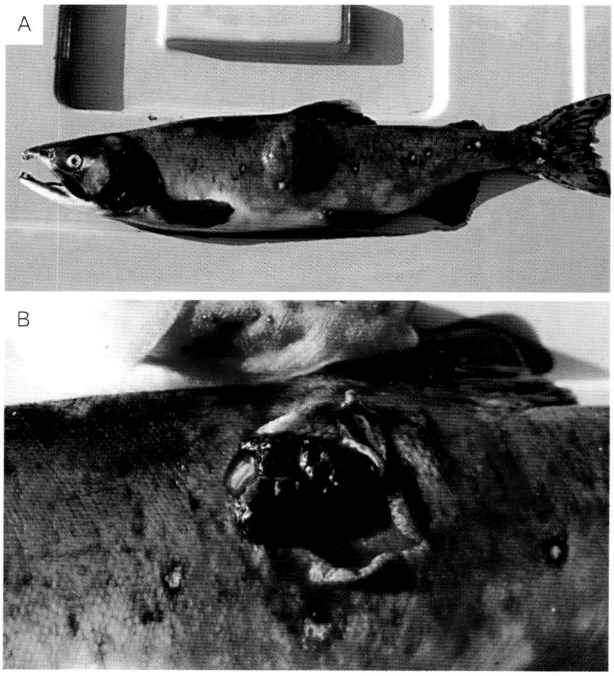

患鱼病变

A．病鱼形成典型的局部脓肿

B．脓肿破溃后形成溃疡

[源自《新鱼病图鉴》，小川和夫]

Macroscopic findings of affected fish

A．Large furuncle is present on the truck of affected fish body

B．Marked ulceration of ruptured furuncle

[Source：*New Atlas of Fish Diseases*，Kazuo Ogawa]

柱状黄杆菌病

疾病概述

【概述】 柱状黄杆菌病由鱼体与病原菌直接接触而引起的细菌性鱼病。

【宿主】 主要感染草鱼（*Ctenopharyngodon idellus*）和青鱼（*Mylopharyngodon piceus*），可人工感染鲤（*Cyprinus carpio carpio*）、黑鲫（*Carassius carassius*）、鲢（*Hypophthalmichthys molitrix*）、鳙（*Aristichthys nobilis*）、团头鲂（*Megalobrama amblycephala*）、金鱼（*Carassius auratus*）等。

【易感阶段】 各种养殖阶段都可发生，尤其是鱼的鳃部遭到机械损伤后更易感染，常造成大批鱼种死亡，常和传染性肠炎、出血病、赤皮病并发。

【发病水温】 发病季节长，该病在水温15℃以上开始流行，水温28 ～ 35℃时为发病高峰期。

【地域分布】 流行于世界各地。

病原

（1）病原为柱状黄杆菌（*Flavobacterium columnaris*），曾用名有鱼害黏球菌（*Myxococcus piscicola*）、柱状屈挠杆菌（*Flexibacter columnaris*）、柱状嗜纤维菌（*Cytophaga columnaris*）。

（2）属黄杆菌科（Flavobacteriaceae）、黄杆菌属（*Flavobacterium*）。

（3）菌体细长、柔软而易弯曲，粗细基本一致，菌体长短很不一致，大多长2 ～ 24μm，个别长37μm，宽0.8μm。两端钝圆，一般稍弯，有时弯成半圆形、圆形、U形、V形或Y形等，以横分裂繁殖，无鞭毛，运动方式为滑行运动和摇晃运动。革兰氏染色阴性，好氧及兼性厌氧菌。

（4）生长最适温度为28℃，37℃时仍可生长，5℃以下则不生长。pH 6.5 ～ 8均可生长，pH 8.5不生长。

（5）可分为基因型Ⅰ、基因型Ⅱ、基因型Ⅲ3种基因型。

临床症状和病理学变化

（1）病鱼体色发黑，尤以头部为甚，游动缓慢，对外界刺激反应迟钝，呼吸困难，食欲减退，鱼体消瘦，常离群独游。

（2）鳃盖内表面的皮肤充血发炎，中间部分有圆形或不规则的糜烂（俗称“开天窗”）。鳃上黏液增多，鳃丝肿胀、缺血而呈淡红色或灰白色，也有呈紫红色，甚至点状出血，严重时鳃小片坏死，鳃丝末端缺损，鳃丝软骨外露，病变鳃丝的周围有淡黄色黏液，带有污泥和杂物碎屑，有时在鳃瓣上可见血斑点。

(3) 病理组织学检查，鳃丝和鳃小片变得软弱，失去张力，往往呈现凋萎不整的弯曲，鳃小片上皮细胞肿大变性，毛细血管充血、渗出，严重时鳃丝呈棍棒状。部分肝细胞发生颗粒变性及水样变性，严重时可见肝细胞内的小空泡连成大空泡，核悬浮在肝细胞的中央，细胞周界不清晰，细胞核肿大，并出现部分肝细胞的核浓缩、破裂、溶解消失。

(4) 肾组织中肾小管的病变最为明显，肾小管水肿，上皮细胞与基底膜完全分离，形成一空腔，还可见到肾小管的上皮细胞增生，细胞排列不规则，将肾小管的管腔堵塞。肾小管有大量的血细胞浸润，较严重的某些肾小管的上皮细胞发生变性，甚至死亡，仅残留核及细胞碎屑。

诊断方法

(1) 细菌分离 使用胰胨琼脂培养基平板，25℃培养2～3d，28～30℃培养24h。菌落黄色，大小不一，有扩散性，中央较厚，显色较深，向四周扩散成颜色浅的假根头。

(2) 理化生化鉴定 革兰氏阴性，氧化酶试验阴性，过氧化氢酶试验阳性，葡萄糖利用产气试验阴性，不分解蔗糖、甘露醇，不产生H_2S，能分解尿素，靛基质试验阴性。

(3) 16S rRNA鉴定 27F (5′-AGA-GTT-TGA-TCM-TGG-CTC-AG-3′) 和1 387R (5′-GGG-CGG-WGT-GTA-CAA-GGC-3′)，退火温度为55℃，扩增产物约为500bp。

扩增产物经测序后判定。

(4) 巢式PCR

第一步引物为：FvpF1 (5′-GCC-CAG-AGA-AAT-TTG-GAT-3′) 和FvpR1 (5′-TGC-GAT-TAC-TAG-CGA-CC-3′)，退火温度为59℃。

第二步引物为：ColF (5′-CAG-TGG-TGA-AAT-CTG-GT-3′) 和ColR (5′-GCT-CCT-ACT-TGC-GTA-GT-3′)，退火温度为48℃，扩增产物长度约为675bp。

扩增产物经测序后判定。

防治方法

(1) 注意保持水体稳定，经常调水、改底，鱼发病后通过外泼消毒剂和内服抗菌药物治疗。

(2) 彻底清塘，鱼池施肥时应施用经过充分发酵后的粪肥。

(3) 鱼种下塘前，用10mg/L的漂白粉溶液或15～20mg/L的高锰酸钾溶液，药浴15～30min；或用2%～4%的食盐溶液药浴5～10min。

(4) 在发病季节，每月全池遍洒生石灰1～2次，使池水的pH保持在8左右（用药量视水的pH而定，一般为15～20mg/L）。

(5) 发病季节，每周在食场周围泼洒漂白粉1～2次，消毒食场。用量视食场大小及水深而定，一般为250～500g。也可用挂篓法预防。

(6) 定期将乌桕叶扎成数小捆，放在池水中浸泡，隔天翻动1次。

（7）含氯消毒剂全池遍洒，以漂白粉（含有效氯25%～30%）1～1.2mg/L的浓度计算用量。

（8）种苗在运输过程中，用1/3 000的硫酸铜溶液处理20min，20～50mg/L的高锰酸钾处理10～15min，或用1/2 000的高锰酸钾处理1～2min。

Columnaris disease

Disease overview

[Disease Characteristic] The disease is acquired from direct contact between fish and pathogens.

[Susceptible Host] Mainly affect grass carp (*Ctenopharyngodon idellus*), black carp (*Mylopharyngodon piceus*), common carp (*Cyprinus carpio carpio*), crucian carp (*Carassius carassius*), silver carp (*Hypophthalmichthys molitrix*), bighead carp (*Aristichthys nobilis*), Wuchang bream (*Megalobrama amblycephala*) and goldfish (*Carassius auratus*), etc.

[Susceptible Stage] Disease affects all stages of fish in aquaculture, in particular when the gill is physically damaged, often resulting in massive death of fish. The diease is often concurrent with infectious enteritis, hemorrhagic disease and red-skin disease.

[Outbreak Water Temperature] Disease outbreaks occur when water temperature is above 15℃, with peaks at 28~35℃.

[Geographical Distribution] Widely distributed globally.

Aetiological agent

(1) *Flavobacterium columnaris*, previously named as *Myxococcus piscicola, Flexibacter columnaris, Cytophaga columnaris*.

(2) Family: Flavobacteriaceae. Genus: *Flavobacterium*.

(3) Gram negative bacteria, aerobic or facultative anaerobic, slender, soft and easily curved, uniform width but the length varies, usually around 0.8μm in width and 2~24μm in length but can be as long as 37μm for some individual. Both ends are obtuse, usually slightly curved, sometimes curved as semicircle, circular, U-shaped, V-shaped or Y-shaped. Reproduce by binary fission, no flagella, with sliding and shivering movement.

(4) Optimal growth at 28℃, able to grow at 37℃ but not below 5℃. Able to grow at pH 6.5~8, but not at pH 8.5.

(5) Subcategorized into genotypes Ⅰ, Ⅱ and Ⅲ.

Clinical signs and pathological changes

(1) Darkened body color, especially on the head, slow movement, delayed response to external stimuli, dyspnea, inappetence and often segregated.

(2) Inflammation and hyperaemia of the inner surface of operculum, with circular or irregular erotic center. Gill with marked increase of mucus, swollen gill filaments, being anaemic and appear pale red or greyish white in color, sometimes purple-red or even with petechiae. In severe case, gill lamella show small areas of necrosis damaging extremities of the filaments and exposing the supporting cartilage. The affected filaments are surrounded by pale yellow mucus, with sludge and debris, sometimes with petechial haemorrhage on the flap.

(3) Histological examination reveals weakened filaments and lamella losing tension, often wilting and bending. Epithelial cells of the lamella are swollen and metaplastic, with hyperaemic and exuded capillaries. In severe cases, the filament appears rob-like. Some of the hepatocytes undergo particular degeneration or hydropic degeneration; small vacuoles joined and become large vacuoles suspending the nuclei at the centre of the cells in more severe cases. Cell margins are unclear with swollen nuclei with pyknosis, karyorrhexis and karyolysis.

(4) The changes of swollen renal tubules are the most prominent in the kidney, the epithelial cells and basement membrane are detached, with renal tubular epithelial hyperplasia, irregular cell arrangement obscuring the lumens. Due to numerous blood cells infiltrated in the renal tubules, severely affected tubules becomes degenerate or death, with cellular debris remnants.

Diagnostic methods

(1) **Bacterial isolation** Use Trypticase soy agar (TSA) plate, incubate for 2~3 days at 25℃ further sub-culture and incubate at 28~30℃ for 24 hours. Bacterial colonies are yellow, variably sized, diffusing, with thick and darkened centre, and swarm to the surrounding with light-colored pseudo-roots.

(2) **Biochemical identification** Gram-negative, oxidase-negative, hydrogen peroxide test positive, gas production glucose utilization gas test negative, non-sucrose and non-mannitol fermenting, hydrogen sulfide production test negative, urease test positive and indole test negative.

(3) **16S rRNA sequencing identification** Perform PCR amplification using the primers 27F (5′-AGA-GTT-TGA-TCM-TGG-CTC-AG-3′) and 1, 387R (5′-GGG-CGG-WGT-GTA-CAA-GGC-3′) with annealing temperature at 55℃ . The amplicon size is 500bp.

Perform sequencing on the amplicon for confirmation.

(4) **Nested-polymerase chain reaction (PCR)**

① First step primers: FvpF1 (5′-GCC-CAG-AGA-AAT-TTG-GAT-3′and FvpR1 (5′-TGC-GAT-TAC-TAG-CGA-CC-3′) with annealing temperature at 59℃ .

② Second step primers: ColF (5′-CAG-TGG-TGA-AAT-CTG-GT-3′ and ColR (5′-GCT-CCT-

ACT-TGC-GTA-GT-3′) with annealing temperature at 48℃ . The amplicon size is 675bp, Perform sequencing on the amplicon for confirmation.

Preventative measures

(1) Maintain stable water condition. Closely adjust water condition accordingly and remove sludge. After disease outbreak, disinfect pond with disinfectant and treat the fish with oral antibiotics.

(2) Clean the pond thoroughly. Fertilize with adequately fermented manure.

(3) Before stocking of fingerlings, use 10mg/L bleaching solution or 15~20mg/L potassium permanganate solution as medicated bath for 15~30 min, or with 2%~4% sodium chloride solution for 5~10 min.

(4) During outbreak seasons, apply quicklime to the pond once or twice a month to maintain the pH of the water at around 8 (dosage depends on the pH of the water, usually around 15~20mg/L).

(5) During outbreak seasons, apply bleaching powder once or twice around the food ponds every week. Dosage depends on the size and water depth of the pond, usually around 250~500g. Prevention can also be done with basketing method

(6) Prepare small bundles of tallow tree leaves (*Sapium sebiferum*) regularly, soak in the pond and flip them once every other day.

(7) Apply chlorine-based disinfectant throughout the pond or bleaching powder (containing 25%~30% of available chlorine) at a concentration of 1~1.2mg/L.

(8) During transportation of fry/fingerling, use 1/3,000 Copper sulfate solution as treatment for 20 minutes, 20~50mg/L potassium permanganate treatment for 10~15 min, or 1/2,000 potassium permanganate for 1~2 min.

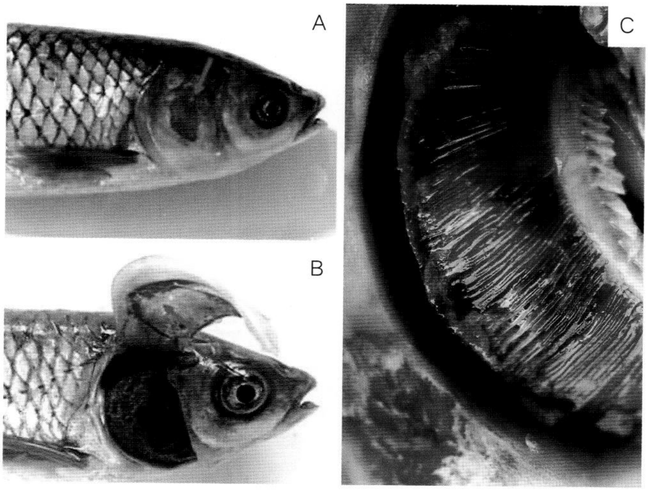

患鱼病变

A．鳃盖被腐蚀　B．鳃盖内表皮充血

C．鳃部感染情况：烂鳃、出血及增生

[源自武汉中博水产；《新鱼病图鉴》，小川和夫]

Macroscopic findings of affected fish

A．Erosion and ulceration of the operculum

B．Haemorrhage present on the internal surface of the operculum. Marked necrosis of the gill towards the tip of the gill filaments C．Closer view of the gill. Marked necrosis, haemorrhage and hyperaemia of the gill filaments

[Source：ZHONGBO AQUATIC，WUHAN；*New Atlas of Fish Diseases*，Kazuo Ogawa]

细菌性冷水病

疾病概述

【概述】 细菌性冷水病是一种对鱼类饲养和保护造成严重影响的疾病，又称尾柄病。也曾被命名为烂鳍病、鱼苗死亡综合征。在欧洲，该病主要指虹鳟苗综合征（RIFS）。

【宿主】 主要感染虹鳟（*Oncorhynchus mykiss*）和银大麻哈鱼（*O. kisutch*）幼鱼以及欧洲鳗鲡（*Anguilla anguilla*）、鲤（*Cyrpinus carpio*）、丁鲹（*Tinca tinca*）等。

【易感阶段】 可感染各个阶段的鱼，幼鱼一旦感染，死亡率达50%以上。

【发病水温】 多发生在4 ~ 10℃，水温15℃时最为严重。水温在10 ~ 16℃时，银鲑的死亡率随着水温升高而增加。一般死亡率为10%~ 30%，最高的死亡率报道达90%。

【地域分布】 1941年在美国发现，现流行于世界各地，包括整个北美洲和几乎所有的欧洲国家以及澳大利亚、智利、秘鲁、日本、韩国、土耳其等。

病原

（1）病原为嗜冷黄杆菌（*Flavobacterium psychrophilum*）。

（2）属黄杆菌科（Flavobacteriaceae）、黄杆菌属（*Flavobacterium*）。

（3）在生长过程中由球杆状变为细杆状，通常为0.5μm×（1.0 ~ 3.0）μm。周身有鞭毛，不形成芽孢，革兰氏染色阴性菌，严格好氧。

（4）一种在低温环境下生存的极端环境微生物，其生长温度范围在4 ~ 25℃，最适生长温度为21℃。

（5）嗜冷黄杆菌有多种不同的血清型和基因型，致病性菌株内部也存在多种不同的变异菌。

临床症状和病理学变化

（1）沿着鳍缘逐渐出现白色糜烂样病变，接着是出现渐进性坏死。皮肤溃疡，呈苍白色，鳃坏死，表皮增生，黏液分泌增加，黑色素沉着增加（“黑尾”现象）。

（2）有腹水，脾脏增大，肠道炎症，神经紊乱，脊柱畸形，肛门出血等。

诊断方法

（1）分离鉴定 使用血琼脂平板，21℃下24h培养，产生半透明、光滑、全缘或偶尔不透明的黄色菌落。

（2）理化生化鉴定 革兰氏阴性，无动力，氧化酶阳性，硝酸盐还原，不产H_2S，不水解淀粉，液化明胶，不产生脲酶，产生脂酶，不发酵葡萄糖。

（3）酶联免疫吸附试验（ELISA）

（4）巢式PCR

第一步引物为：20F（5′-AGA-GTT-TGA-TCA-TGG-CTC-AG-3′）和1 500R（5′-GGT-TAC-CTT-GTT-ACG-ACT-T-3′），退火温度为57℃。

第二步引物为：PSY1（5′-GTT-GGC-ATC-AAC-ACA-CT-3′）和PSY2（5′-CGA-TCC-TAC-TTG-CGT-AG-3′），退火温度为57℃，扩增产物长度为1 089 bp。扩增产物进行测序后判定。

（5）Real-time PCR 引物为FpSigF（5′- GGT-AGC-GGA-ACC-GGA-AAT-G-3′）和FpSig R（5′-TTT-CTG-CCA-CCT-AGC-GAA-TAC-C-3′），探针FpSig（Tet-5′-CGC-TTC-CTG-AGC-CAG-A-MGBNFQ-3′），退火温度为60℃。

防治方法

（1）尽可能地减少鱼类的应激，去除体外寄生虫，确保良好的水质，提供良好的营养，做好生物安保工作。

（2）土霉素是世界各地防控细菌性冷水病常见的药物。不过，嗜冷黄杆菌的抗药性已出现，因此，抗生素疗法是一种不可持续的防控方法。

（3）可在鱼卵的水硬化过程中，使用吡咯烷酮碘100mg/L持续10min，或50mg/L持续30min，可以明显降低细菌的数量。

（4）处于试验阶段的一种福尔马林灭活或高温灭活菌苗用于注射接种表现出了一些效果，但无法用于大规模的鱼苗免疫接种。

Bacterial cold water disease, BCWD

Disease overview

[Disease Characteristic] A disease having severe impact on the production and protection of fish. The disease was initially named as peduncle disease. It has also been named as rotten fin disease and fry death syndrome. In Europe, the disease is mainly called rainbow trout fry syndrome (RTFS).

[Susceptible Host] Mainly affect juveniles of rainbow trout (*Oncorhynchus mykiss*) and silver salmon (*O. kisutch*), European eel (*Anguilla anguilla*), common carp (*Cyprinus carpio*), tench (*Tinca tinca*), etc.

[Susceptible Stage] Fish of any age can be affected. Mortality rate may reach over 50%

in juveniles.

[Outbreak Water Temperature] Bacterial cold water disease outbreaks occur mostly at 4~10℃ , with most severe outbreaks occurring at 15℃ . At water temperature between 10~16℃ , mortality of silver salmon increases as the water temperature increases. Mortality rate is usually around 10~30%, with the highest ever reported as 90%.

[Geographical Distribution] First reported in the USA in 1941, and is now endemic globally including the entire North America and almost all European countries, Australia, Chile, Peru, Japan, South Korea, Turkey, etc.

Aetiological agent

(1) *Flavobacterium psychrophilum.*

(2) Family: Flavobacteriaceae. Genus: *Flavobacterium*;

(3) Shape changes from cocci to small rod during growth, usually 0.5μm × (1.0~3.0)μm in size with flagella around the body, non-spore forming. Gram-negative bacteria, obligate aerobe.

(4) Extremophile that can survive in low temperature environments, with growing temperatures ranging from 4℃ to 25℃ and optimal growth at 21℃.

(5) Many different serotypes and genotypes. There are also many variants among the pathogenic strains.

Clinical signs and pathological changes

(1) White erosion-like lesions gradually appear along the edge of the fin, followed by progressive necrosis. Skin ulcerations and paling. Gill necrosis, epidermal hyperplasia, increased mucus secretion, melanosis (causing "black tail").

(2) Ascites, splenomegaly, enteritis, neurological disorders, spinal deformity, anal bleeding, etc.

Diagnostic methods

(1) **Bacterial isolation** Incubate on blood agar at 21℃ for 24 hours. The bacteria form yellow colonies which are typically translucent, smooth, round, and occasionally opaque.

(2) **Biochemical identification** Gram-negative, non-motile, oxidase-positive, nitrate reducing, non-H_2S producing, not hydrolyzing starch and gelatin, non-urease producing, non-lipase producing, and non-glucose-fermenting.

(3) **Enzyme-linked immunosorbent assay (ELISA)**

(4) **Nested-polymerase chain reaction (PCR)**

First step primers: 20F (5′-AGA-GTT-TGA-TCA-TGG-CTC-AG-3′) and 1,500R (5′-GGT-TAC-CTT-GTT-ACG-ACT-T-3′) with annealing temperature at 57℃.

Second step primers: PSY1 (5′-GTT-GGC-ATC-AAC-ACA-CT-3′) and PSY2 (5′-CGA-TCC-TAC-TTG-CGT-AG-3′) with annealing temperature at 57℃. The amplicon size is 1,089bp. Perform sequencing on the amplicon for confirmation.

(5) **Real-time PCR** Use the primers:

FpSig F (5′- GGT-AGC-GGA-ACC-GGA-AAT-G-3′) and FpSig R (5′-TTT-CTG-CCA-CCT-AGC-GAA-TAC-C-3′) with FpSig probe (Tet-5′-CGC-TTC-CTG-AGC-CAG-A-MGBNFQ-3′). The annealing temperature is 60℃.

Preventive measures

(1) Reduce the stress of the fish as much as possible, including eradicating ectoparasites, ensuring good water quality, providing good nutrition and maintaining good biosecurity.

(2) Oxytetracycline is a common antibiotic used for prevention of bacterial cold water disease worldwide, yet antimicrobial resistance has already been noted in *F. psychrophilum*. Thus, antibiotic treatment would not be a sustainable preventive measure.

(3) Pyrrolidone iodine: 100mg/L for 10 minutes or 50mg/L for 30 minutes during the hardening process of fish eggs can significantly reduce the amount of bacteria.

(4) Formalin-inactivated or heat-inactivated vaccine in the experimental phase showed some effectiveness by injection. However, it cannot be used for large-scale fry immunization.

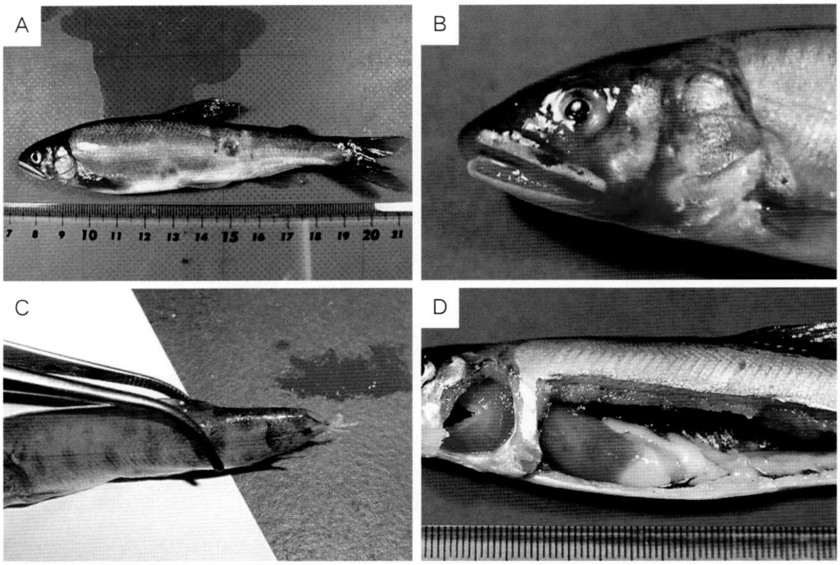

患鱼病变

A．体表糜烂、溃疡　B．下颌部出血　C．鲑科鱼类稚鱼发病的特征为尾柄部糜烂、溃疡以及缺损等　D．鳃及内脏各器官贫血

[源自《新鱼病图鉴》，小川和夫]

Macroscopic findings of affected fish

A．Multifocal erosions and ulcerations on the body trunk　B．Submandibular haemorrhage

C．Erosions, ulcerations, or damage to the tail of Salmonidae juvenile fish

D．Anaemia of gill and other internal organs

[Source：*New Atlas of Fish Diseases*，Kazwo Ogawa]

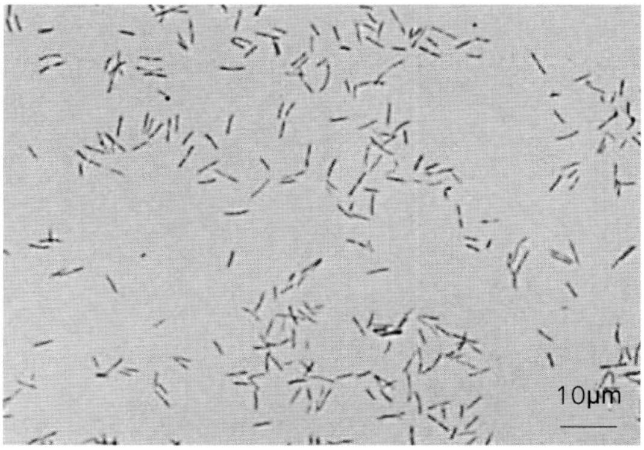

病原菌为革兰氏阴性的嗜冷黄杆菌

(*Flavobacterium psychrophilum*)

大小为（2~7）μm×（0.3~0.75）μm (Gram染色)

[源自《新鱼病图鉴》，小川和夫]

The aetiological agent is Gram-negative *Flavobacterium psychrophilum*

(2~7) μm × (0.3~0.75) μm in size (Gram staining)

[Source：*New Atlas of Fish Diseases*，Kazuo Ogawa]

链球菌病

疾病概述

【概述】 链球菌病是一种广泛危害淡水、海水养殖鱼类的感染宿主广、传染性强、死亡率高的细菌性疾病，给水产养殖业特别是集约化鱼类养殖造成了巨大的经济损失。

【易感宿主】 主要感染海水养殖的杜氏鰤（*Seriola dumerili*）、日本竹筴鱼（*Trachurus japonica*）、缟鲹（*Caranx delicatissimus*）、条石鲷（*Oplegnathus fasciatus*）、牙鲆（*Paralichthys olevaceus*）、真鲷（*Chrysophrys major*）、黑鲷（*Sparus macrocephalus*）、红鳍东方鲀（*Fugu rubripes*）和半咸水及淡水养殖的日本鳗鲡（*Anguilla japonica*）、香鱼（*Plecoglossus altivelis*）、虹鳟（*Salmo gairdneri*）及尼罗罗非鱼（*Tilapia nilotica*）等。

【易感阶段】 感染各种大小的鱼，从当年鱼种至成鱼均可受感染，但较大的鱼（大于100g）更易感。

【发病水温】 全年大部分时间可以发生，水温为25 ~ 37℃时是流行高峰，高于30℃时易导致疾病暴发。疾病常呈急性型，在高水温季节时，病鱼的死亡高峰期可持续2 ~ 3周；但在低水温季节，该病也可呈慢性型，死亡率较低，但持续时间长。

【地域分布】 北美洲、南美洲、欧洲、亚洲等，如日本、以色列、美国、科威特、泰国、中国、越南、巴西等。

病原

（1）病原有海豚链球菌（*Streptococcus iniae*，曾用名 *Streptococcus shiloi*）、无乳链球菌（*S. agalactiae*）、副乳房链球菌（*Streptococcus parauberis*）、格氏链球菌（*Lactococcus garvieae*）等，主要是海豚链球菌和无乳链球菌。

（2）链球菌科（Strertococcaceae）、链球菌属（*Streptococcus*）。

（3）菌体椭圆形，二链或链锁状的球菌，直径大小为0.2 ~ 0.8μm，有荚膜，无鞭毛，不形成芽孢，无运动能力，革兰氏阳性菌，有时老龄菌呈革兰氏阴性，兼性厌氧。

（4）菌落呈链状排列，链长短与培养基和菌株有关，生长温度为10 ~ 45℃，最适温度为20 ~ 37℃。生长盐度范围为0 ~ 70，最适盐度为0。生长pH为3.5 ~ 10，最适pH为7.6。

（5）并非所有海豚链球菌都是致病菌，存在不致病的共生菌株。致病菌细胞壁中的类M蛋白是一种毒力因子，能抗吞噬作用以及抗吞噬细胞内的杀菌作用，将这些致病菌归为A群链球菌。

（6）根据无乳链球菌表面荚膜多糖组分的差异，可分为10个血清型。其中，Ⅰa、Ⅰb和Ⅲ型为目前确定的水生动物源血清型，均包含*cpsF*基因，且同源性为98%～ 100%。

临床症状和病理学变化

（1）失去食欲，静止于水底，或离群独自漫游于水面，有时做旋转游泳后再沉入水底，鱼体弯曲呈C形或逗号样。

（2）体色发黑，吻端发红，肌肉充血，体表黏液增多，眼球突出，眼球混浊，其周围充血，鳃盖内侧发红、充血或强烈出血。

（3）水温较低时，还会出现各鳍均发红、充血或溃烂，体表局部特别是尾柄往往溃烂或带有脓血的疖疮。

（4）幽门垂、肝脏、脾脏、肾脏或肠管均有点状出血。肝脏因出血和脂肪变形而褪色，甚至组织破损。

（5）肾脏肾小球及褐色素巨噬细胞中心以及肾小管间质组织侵染细菌并大量繁殖，引起肾脏肿大、坏死。中肠道上皮的固有层破损引起肠炎，革兰氏染色可见肠绒毛基部有聚集的菌落。

诊断方法

（1）显微镜检　取病灶组织或鳃盖内侧膜内积液，革兰氏染色，置于油镜下检查，可发现革兰氏阳性的链状球菌，发现较多为成对球菌，可确诊为疑似链球菌病。

（2）生化鉴定　可根据《伯杰氏手册》或API细菌鉴定系统、全自动微生物鉴定系统进行生化鉴定，主要生化特性为革兰氏染色阳性、过氧化氢酶阴性，无运动性，万古霉素阳性，V-P反应阴性，胆汁七叶苷阴性，山梨醇产酸，MRS液体培养基产气。

（3）间接免疫荧光试验（IFAT）

（4）PCR

无乳链球菌上下游引物分别为H1(5′-AAG-CGT-GTA-TTC-CAG-ATT-TCC-T-3′)和H2(5′-CAG-TAA-TCA-AGC-CCA-GCA-A-3′)，退火温度为58℃，扩增产物长度为474 bp。

海豚链球菌上下游引物有两组，一组为P1（5′-CTA-GAG-TAC-ACA-TGT-ACT-TAA-G-3′）和P2（5′-GGA-TTT-TCC-ACT-CCC-ATT-AC-3′），退火温度为58℃，扩增产物长度为296bp；另一组为LOX-1（5′-AAG-GGG-AAA-TCG-CAA-GTG-CC-3′）和LOX-2（5′-ATA-TCT-GAT-TGG-GCC-GTC-TAA-3′），退火温度为55℃，扩增产物长度为870bp。

扩增产物经测序后判定。

（5）16S rDNA鉴定　P_{LS}（5′-AGA-GTT-TGA-TCC-TGG-CTC-AG-3′）和P_{LA}（5′-TAC-GGC-TAC-CTT-GTT-ACG-ACT-T-3′），退火温度为57℃，扩增产物长度为1 440bp。

扩增产物经测序判定。

防治方法

（1）保持良好的水质状况和合理的放养密度，在高温季节应适当提高池塘水位，饵料必须新鲜，勿过量投饲，增大水体交换，增加水体的溶解氧。

（2）用生石灰、含氯消毒剂进行水体的消毒。

（3）在饲料中添加维生素C、免疫多糖、酵母细胞壁等免疫增强剂。

（4）用盐酸强力霉素制成药饵，连续用药14d以上。

（5）博落回、紫草、田基黄、补骨脂、三颗针、田七须和五倍子，对无乳链球菌和海豚链球菌均具有较好的抑菌效果。

（6）腹腔接种灭活疫苗，相对保护率达93.2%。目前，还研发出二联疫苗、减毒疫苗、亚单位疫苗和DNA疫苗等。

Streptococcosis

Disease overview

[Disease Characteristic] Widespread bacterial disease affecting a large variety of freshwater and marine fish species. Highly contagious with high mortality rate, causing huge economic losses to aquaculture industry, especially to fish farms with intensive aquaculture practices.

[Susceptible Host] Mainly affect marine fish aquaculture including greater amberjack (*Seriola dumerili*), Japanese horse mackerel (*Trachurus japonica*), white trevally (*Caranx delicatissimus*), striped beakfish(*Oplegnathus fasciatus*), olive flounder (*Paralichthys olevaceus*), red seabream (*Chrysophrys major*), black seabream (*Sparus macrocephalus*), Takifugu rubripes (*Fugu rubripes*) as well as fresh water and brackish water aquacultures such as Japanese eel (*Anguilla japonica*), sweetfish (*Plecoglossus altivelis*), rainbow trout (*Salmo gairdneri*), Nile tilapia (*Tilapia nilotica*), etc.

[Susceptible Stage] Infection can occur in any stages from fingerling to adult, but larger fish (larger than 100g) are more susceptible to infection.

[Outbreak Water Temperature] Disease outbreaks occur all year round, being endemic at water temperature of 25~37℃ , outbreaks tend to occur above 30℃. Commonly acute onset of disease. During seasons of high water temperature, peak mortality can last for 2~3 weeks. However, in seasons of low water temperature, the disease can also persist as chronic infection with low mortality rate for a long time.

[Geographical Distribution] Disease distributed in North America, South America, Europe and Asia, such as Japan, Israel, the United States, Kuwait, Thailand, China, Vietnam, Brazil, etc.

Aetiological agents

(1) Mainly *Streptococcus iniae* (previously named *Streptococcus shiloi*) and *S. agalactiae*. Others include *S. parauberis*, *Lactococcus garviea* etc.

(2) Family: Streptococcaceae. Genus: *Streptococcus*.

(3) Oval bacteria. Gram-positive (may become Gram-negative if aged), facultative anaerobic, paired or chain-like cocci, around 0.2μm × 0.8μm in size, with capsule, without flagella, non-sporulated and non-motile.

(4) Bacterial colonies are arranged in chains, and the length of the chains is related to the culture medium and the bacterial strain. Bacterial growth at temperature ranges from 10℃ to 45℃, optimal growth at 20~37℃; salinity ranges from 0 to 70, optimal salinity at 0; pH ranges from 3.5 to 10, optimal pH at 7.6.

(5) *Streptococcus iniae* may also exist as a commensal that not all strains of are considered pathogenic. The pathogenicity is related to a virulent factor named M-like protein (emm), which is a component protein in the cell wall of the streptococcus with anti-phagocytic functions. These pathogenic *Streptococci* are grouped as Group A streptococcus.

(6) For *Streptococcus agalactiae*, it can be sub-divided into 10 serotypes according to different components of the capsular polysaccharide. Ⅰa, Ⅰb and Ⅲ are currently identified as serotypes originated from aquatic animal which all contain the *cpsF* gene, with 98% ~100% homology.

Clinical signs and pathological changes

(1) Inappetence, stay still at the bottom of the pond or segregated near the surface of the water, sometimes sink to the bottom of the pond after whirling. Affected fish bending the body in a "C" or comma shape.

(2) Darkened body color, reddening of the tip of the mouth, hyperaemia of muscles hyperaemia, increased mucus secretion on body surface, exophthalmos, turbid eyeballs with periocular hyperaemia, and reddening of the inner operculum, being hyperaemic or highly hemorrhagic.

(3) Under lower water temperature, all the fins are reddened, hyperaemic or necrotic. Part of the body surface, especially the base of tail, usually ulcerated or with furuncles of purulent discharge.

(4) Petechiae in the pylorus, liver, spleen, kidney or intestines. Liver discoloration due to fatty changes and haemorrhages; may even cause tissue damages.

(5) Renal glomeruli, melano-macrophages centres and interstitial tissues are invaded with bacteria. Bacterial propagation causes renal enlargement and necrosis. Damage to the lamina propria of the midgut epithelium causes enteritis, bacterial colonies can be revealed at the base of the intestinal villi by Gram staining.

Diagnostic methods

(1) **Microscopic examination** Collect tissues from relevant lesions or fluid accumulated in the operculum as samples. Perform Gram stain and examine under microscope with oil immersion. Look for Gram-positive streptococci, it can be suspected as Streptococcosis if predominately paired cocci are noted.

(2) **Biochemical identification** With reference to *Bergey's Manual* or using API identification system or automatic microbial identification system to conduct microbiology biochemical identification. Major biochemical characteristics are: gram-positive, hydrogen peroxide test negative, non-motile, sensitive to vancomycin, negative for V-P reaction, no growth on bile esculin agar, acid production in sorbitol test, gas production in MRS liquid medium, etc.

(3) **Indirect fluorescent antibody technology (IFAT)**

(4) **Polymerase chain reaction (PCR)**

Streptococcus agalactiae: The primers are H1 (5′-AAG-CGT-GTA-TTC-CAG-ATT-TCC-T-3′) and H2 (5′-CAG-TAA-TCA-AGC-CCA-GCA-A-3′) with annealing temperature at 58 ℃. The amplicon size is 474 bp.

Streptococcus iniae: The primers are P1 (5′-CTA-GAG-TAC-ACA-TGT-ACT-TAA-G-3′) and P2 (5′-GGA-TTT-TCC-ACT-CCC-ATT-AC-3′) with annealing temperature at 58℃. The amplicon size is 296bp.

Streptococcus iniae: The primers are LOX-1 (5′-AAG-GGG-AAA-TCG-CAA-GTG-CC-3′) and LOX-2 (5′-ATA-TCT-GAT-TGG-GCC-GTC-TAA-3′) with annealing temperature at 55℃. The amplicon size is 870bp.

Perform sequencing on the amplicons for confirmation.

(5) **16S rDNA identification** Perform PCR amplification using the primers P_{LS} (5′-AGA-GTT-TGA-TCC-TGG-CTC-AG-3′) and P_{LA} (5′-TAC-GGC-TAC-CTT-GTT-ACG-ACT-T-3′) with annealing temperature at 57℃. The amplicon size is 1,440bp. Perform sequencing on the amplicon for confirmation.

Preventative measures

(1) Maintain good water quality and reasonable culture population density. In hot seasons, raise the water level as appropriate. Use fresh feed and do not overfeed. Improve water exchange and increase dissolved oxygen level in water.

(2) Disinfect water with quicklime and chlorinated disinfectant.

(3) Supplement vitamin C, polysaccharide, enzymes or/and other immune boosters in feed.

(4) Prepare medicated feed by mixing doxycycline hydrochloride in feed and treat consecutively for more than 14 days.

(5) Boluohui (*Macleaya cordata*), gromwell (*Lithospermum erythrorhizon*), Tianjihuang

(*Grangea maderaspatana*), psoralen (*Psoralea corylifolia* L.), Barberry root (*Berberis pliretii* Schneid), Chinese ginseng (*Panax notoginseng*) and Galla chinensis (*Rhus chinensis* Mill) have good antibacterial effects against *Streptococcus agalactiae* and *Streptococcus iniae*.

（6）Vaccination: relative protection rate (RPS) of intraperitoneal injection can reach 93.2%. Other vaccines such as bivalent vaccine, attenuated vaccine, subunit vaccine and DNA vaccine are also available.

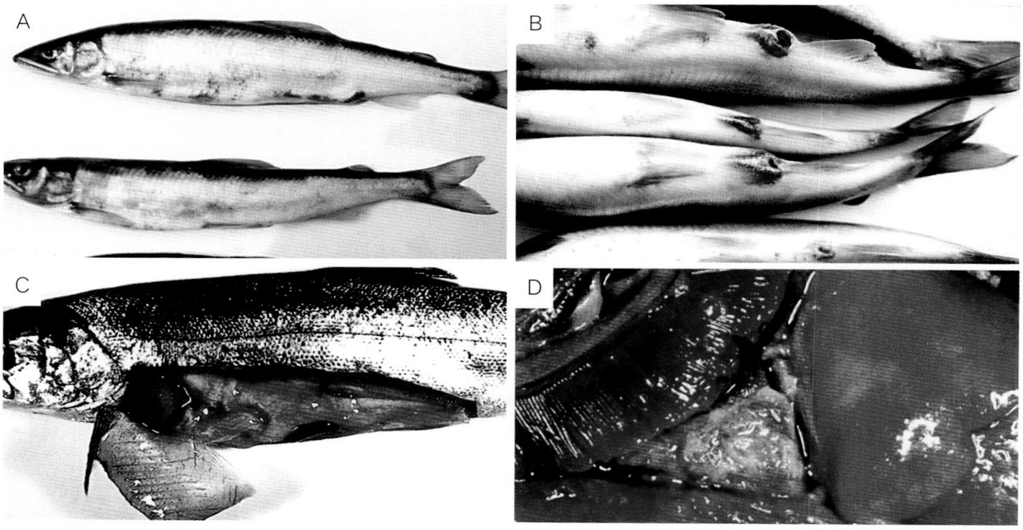

患鱼病变

A．感染海豚链球菌的病鱼眼球突出及眼球内出血等眼球异常现象，鳃盖发红、充血，鱼尾基部溃疡，腹部有点状出血

B．感染海豚链球菌的病鱼肛门周围发红、隆起、开口

C．感染海豚链球菌的病鱼内部器官的病变特征为肠管炎症，脾脏及腹腔内壁出血等

D．感染无乳链球菌的鱼体解剖检查时，特征为心外膜炎

[源自《新鱼病图鉴》，小川和夫]

Macroscopic findings of affected fish

A．Fish infected by *S. iniae*. Exophthalmia and intraocular haemorrhage, hyperaemia of the operculum, ulceration of the tail base and ecchymosis present on the ventral body trunk

B．Fish infected by *S. iniae*. Anal hyperaemia and protrusion

C．Fish infected by *S. iniae*. Visceral haemorrhage and enteritis

D．Fish infected by *S. agalactiae*. Pericarditis is a characteristic lesion

[Source：*New Atlas of Fish Diseases*，Kazuo Ogawa]

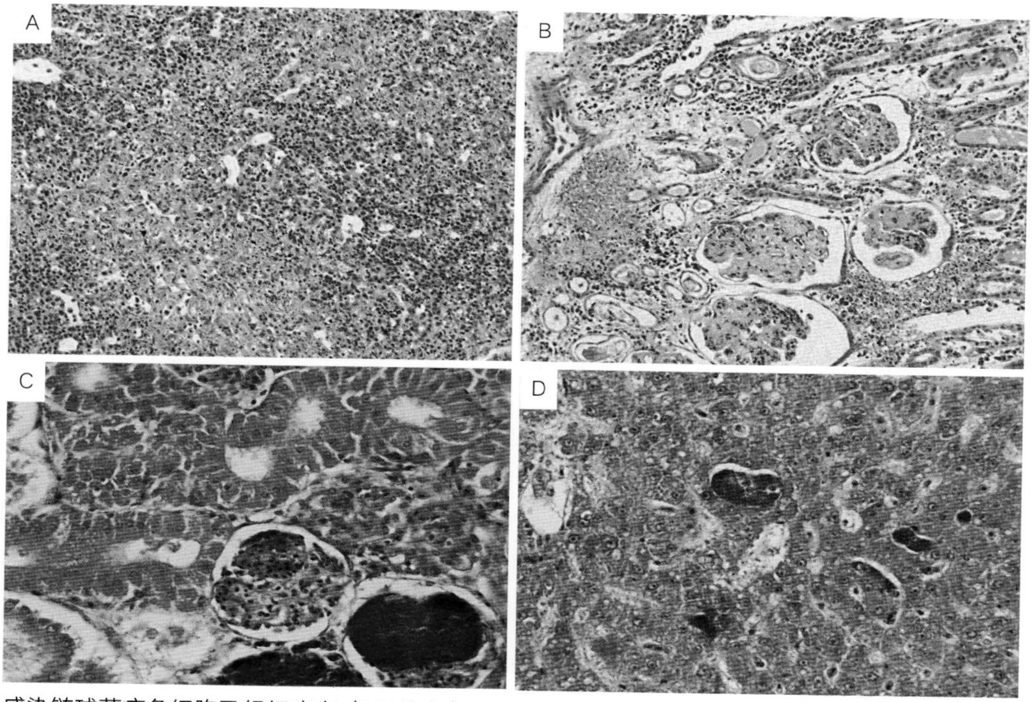

感染链球菌病鱼细胞及组织病变（HE染色）

A．脾脏严重坏死

B．肾脏有坏死灶

C．肾小管萎缩、与基底膜脱落、瘀血并见细菌菌落（蓝染物）

D．肝脏水变性、有蓝染物、肝细胞界限不清晰

[源自雷燕及刘宝芹]

Histopathology of Streptococcosis in fish(HE staining)

Multifocal necrosis in the (A) spleen and (B) kidney

C．Kidney. Tubular degeneration with bacterial colonies present in the section (blue staining)

D．Liver. Mild hydropic degeneration of hepatocytes with bacterial colonies present in the section

[Source：Yan Lei and Baoqin Liu]

赤 皮 病

疾病概述

【概述】 赤皮病又称出血性腐败病、赤皮瘟、擦皮瘟等，是草鱼、青鱼的主要疾病之一。

【宿主】 主要感染草鱼（*Ctenopharyngodon idellus*）、青鱼（*Mylopharyngodon piceus*）、鲤（*Cyprinus carpio*）、团头鲂（*Megalobrama amblycephala*）等多种淡水鱼。

【易感阶段】 多发生于2 ～ 3龄大鱼，当年鱼种也可发生，常与肠炎病、烂鳃病同时发生形成并发症。

【发病水温】 水温25 ～ 30℃时流行；在捕捞、运输后及北方越冬后，最易引发流行。

【地域分布】 全世界均有分布。

病原

（1）病原为荧光假单胞菌（*Pseudomonas fluorescens*）。

（2）属假单胞菌科（Pseudomonadaceae）、假单胞菌属（*Pseudomonas*）。

（3）短杆状，两端圆形，大小为（0.7 ～ 0.75）μm×（0.4 ～ 0.45）μm，单个或两个相连，极端着生1 ～ 3根鞭毛，无芽孢，有运动能力，革兰氏阴性菌，化能异养型的条件致病菌。鱼的体表完整无损时，病原菌无法侵入鱼体，只有当鱼因捕捞、运输、放养造成鱼体受机械损伤、冻伤或体表被寄生虫寄生而受损时，病原菌才能引起发病。

（4）最适生长温度为25 ～ 30℃，大多数菌株能在4℃生长，37 ～ 42℃则不生长。

（5）荧光假单胞菌属于假单胞菌中的rRNA同源群Ⅰ型荧光假单胞菌DNA同源群。在伯杰氏鉴定手册中，依据菌落形态、色素、生理生化和营养的特性等表型特征，将荧光假单胞菌分成5个生物型，分别是生物型Ⅰ、Ⅱ、Ⅲ、Ⅳ、G。

临床症状和病理学变化

（1）病鱼行动缓慢，反应迟钝，衰弱地独游于水面。

（2）体表局部或大部出血发炎，全身肿胀，呈充血发炎的红斑块和化脓性溃疡。鳞片脱落，特别是鱼体两侧和腹部最为明显。

（3）鳍的基部或整个鳍充血，鳍的末端腐烂，常烂出一段，鳍条间的组织也被破坏，使鳍条呈扫帚状，形成“蛀鳍”，或像破烂的纸扇状。在鳞片脱落和鳍条腐烂处，往往出现水霉菌寄生。

（4）鱼的上下颌及鳃盖部分充血，呈块状红斑。鳃盖中部表皮有时烂去一块，以致透明呈小圆窗状。有时鱼的肠道也充血发炎。

（5）腹水增多，肝脏肿大有出血点，肠组织糜烂、溃疡，各器官出血性坏死。

（6）组织病理切片可见细胞出现肿胀、颗粒变形、玻璃样变和坏死崩解，大量红细胞变形、碎裂溶解，呈溶血性贫血，血液中白细胞极少，无白细胞浸润现象。

诊断方法

（1）细菌分离 使用普通营养琼脂平板，28℃下培养20 h左右开始产生绿色或黄绿色色素，弥漫培养基。肉汤培养，生长丰盛，均匀混浊，微有絮状沉淀，表面有光滑柔软的层状菌膜，一摇即碎，24h培养基表层产生色素。马铃薯上中等生长，光滑湿润，菌苔呈绿色。

（2）生化鉴定 可根据《伯杰氏手册》或API细菌鉴定系统、全自动微生物鉴定系统进行生化鉴定，主要生化特性为革兰氏阴性，有运动能力，氧化酶、肌醇、荧光素、明胶均为阳性，过氧化氢酶阳性，吲哚反应阴性，葡萄糖利用产气试验阴性，甘露醇、靛基质阴性。

（3）16S rRNA鉴定 P1（5′-AGA-GTT-TGA-TC（C/A）-TGG-CTC-AG-3′）和P2（5′-GGT-TAC-CTT-GTT-ACG-ACT-T-3′），退火温度为55℃，扩增产物长度为1 455bp。

扩增产物经测序后判定。

预防措施

（1）在捕捞、运输、放养等操作过程中，尽量避免鱼体受伤。鱼种放养前，可用3%～4%的食盐溶液浸洗5～15min或5～8mg/L的漂白粉溶液浸洗20～30min。药浴时间的长短，视水温和鱼体忍受力而灵活掌握。

（2）每隔10～15d，水体用“鱼虾安”（主要成分：三氯异氰脲酸）消毒1次，同时在饵料中添加10g/kg的“鱼病康”（主要成分：恩诺沙星），连续投喂2d，并定期更换新水。

治疗方法

（1）将水温升至26℃，可改善死亡率。

（2）用0.2～0.3mg/L的二氯海因，全池泼洒，在疾病流行季节每15d 1次。

（3）每千克鱼体重每天用四环素40～80mg，或氟苯尼考5～15mg，拌饲投喂，连用3～5d。

（4）磺胺嘧啶饲料投喂，第一天用量是每千克鱼体重用药100mg，以后每天用药50mg，连喂1周。方法是把磺胺嘧啶拌在适量面糊内，然后与草料拌和，稍干一下投喂草鱼。青鱼可拌在米糠或豆饼中投喂。

Red-skin disease

Disease overview

[Disease Characteristic] One of the major diseases of grass carp and black carp, also called bacterial haemorrhagic septicaemia.

[Susceptible Host] Mainly affect freshwater fish species such as grass carp (*Ctenopharyngodon idellus*), black carp (*Mylopharyngodon piceus*), common carp (*Cyprinus carpio*), Wuchang bream (*Megalobrama amblycephala*), etc.

[Susceptible Stage] Usually affects fish at 2~3 years of age, but it can also affect fish of that year. It commonly occurs concurrently with enteritis and gill rot disease to cause complications.

[Outbreak Water Temperature] Endemic throughout the year when water temperature is at 25~30℃. Disease outbreaks particularly common after capturing, transporting and after winter in Northern areas.

[Geographical Distribution] Widely distribution globally.

Aetiological agent

(1) *Pseudomonas fluorescens*.

(2) Family: Pseudomonadacea. Genus: *Pseudomonas*.

(3) Gram-negative short rod bacteria, obtuse at both ends, 0.7~0.75μm × 0.4~0.45μm in size, single or in pairs, with 1~3 polar flagella at the poles, non-spore forming, motile, chemoheterotrophic opportunistic pathogen. The bacteria cannot invade when the fish body is intact. It can only cause disease when the fish body is injured by capturing, transporting, mechanical damage during stocking, cold sores, ectoparasite infections, etc.

(4) The optimal growth temperature is 25~35℃. Most strains are able to grow at 4℃, but cannot survive at 37~42℃.

(5) *P. fluorescens* belongs to type I rRNA homology group in *Pseudomonas*, *P. fluorescens* DNA homology group. In the *Bergey's Manual*, it can be categorized into 5 biotypes (Ⅰ, Ⅱ, Ⅲ, Ⅳ, G) according to the colony morphology, pigment, biochemical and nutritional characteristics, etc.

Clinical signs and pathological changes

(1) Affected fish become inactive, with delayed response and weakness that some may swim

near the water surface alone.

(2) Localized or extensive haemorrhage and inflammation on the body surface. Generalized swelling of body, with congested and inflamed red patches and suppurative ulcers. Detachment of scales, particularly obvious at the sides of the body and the abdomen.

(3) Hyperaemia of the base of the fin or the whole fin. Necrosis of the tip of the fins, a section of necrotic fin is commonly present. Tissues between the fin rays are commonly damaged, distorting the normal architecture of the fin ray to become broom shaped, forming "rot fin" , or resembling a broken paper fan. Secondary *Saprolegnia* infection is common at the location of scale loss and fin necrosis.

(4) Hyperaemia at the upper jaw, lower jaw and the operculum forming red plaques. Necrosis is sometimes noted on the epithelium of the centre of the operculum forming a transparent small round window at the necrotic area. Hyperaemia and inflammation of the intestines are also sometimes noted.

(5) Increased ascites, hepatomegaly with petechiae, intestinal necrosis and ulceration, and multiorgan haemorrhagic necrosis.

(6) Histologically, there are cellular swelling, granulation, hyalinization, necrosis and disintegration. Large number of deformed and fragmented red blood cells are present, causing haemolytic anaemia. Leukopenia resulting in no infiltration of white blood cells.

Diagnostic methods

(1) **Bacterial isolation** Use common nutrient agar and incubate at 28 ℃. Green or yellowish green pigment are produced at around 20 hours of growth and diffuse all over the agar plate. Heavy growth in broth, evenly turbid with mild flocculation, and smooth and soft layer of biofilm on the surface, which is easily broken by shaking. Pigment is produced on the surface of the broth after 24 hours of incubation. Moderate growth on potatoes, being smooth and wet with green bacterial lawn.

(2) **Biochemical identification** With reference to *Bergey's Manual* or using API bacterial identification system or automatic microbial identification system to conduct microbiology biochemical identification tests. The major characteristics are: Gram-negative, motile, positive for oxidase, inositol, fluorescein, gelatin and peroxidase, while negative for indole reaction, glucose fermentation and mannitol fermentation.

(3) **16S rRNA sequencing identification** Use the primers: P1 (5′-AGA-GTT-TGA-TC(C/A)-TGG-CTC-AG-3′) and P2 (5′-GGT-TAC-CTT-GTT-ACG-ACT-T-3′) with annealing temperature at 55℃. The amplicon size is 1,455bp. Perform sequencing on the amplicon for confirmation.

Preventive measures

(1) During capture, transport and stocking, avoid traumatizing the fish. Before releasing the fingerlings into the culture pond, they can be treated with immersion in 3%~4% sodium chloride solution for 5~15 min or solution with 5~8mg/L bleaching powder solution for 20~30 min. The duration of medicated bath depends on the water temperature and the tolerance of the fish, which needs to be monitored carefully.

(2) Disinfect the water with trichloroisocyanuric acid once every 10~15 days. In the meantime, apply 10g/kg enrofloxacin in feed and treat for 2 consecutive days. Also, refill new water regularly.

Treatment

(1) Raise of the water temperature to 26℃ can reduce the mortality rate.

(2) Apply 0.2~0.3mg/L dichlorohydantoin to the whole pond once every 15 days during outbreak seasons of the disease.

(3) For every one kilogram of fish, apply 40~80mg tetracycline, or 5~15mg florfenicol in feed and treat for 3~5 consecutive days.

(4) Sulfadiazine (mix with feed): at dosage of 100mg/kg of fish for the first day, followed by 50mg/kg daily consecutively for a week. The method is to mix sulfadiazine with an appropriate amount of batter, then mix with forage and slightly dried before feeding grass carp. Black carp can be fed by mixing with rice bran or soy cake.

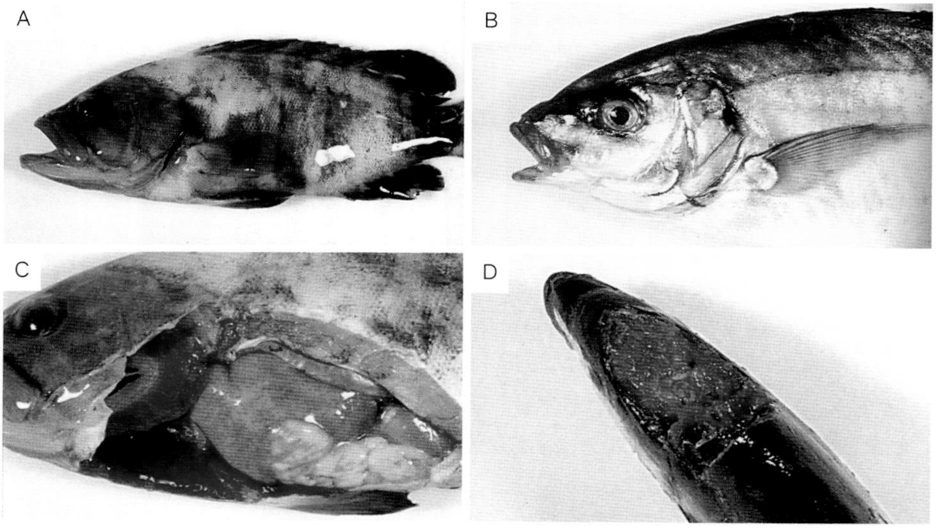

患鱼病变

A. 体表出现伴随脱鳞的擦伤样症状，可见出血 B. 吻部、鳃盖外侧发红和出血

C. 鳃充血，肝脏轻度出血 D. 多数患病鱼体可见脑部发红

[源自《新鱼病图鉴》，小川和夫]

Macroscopic findings of affected fish

A. Massive erosion with scale loss and haemorrhage on the body trunk

B. Hyperaemia and haemorrhage on the mouth and the operculum

C. Hyperaemia of gill and mild petechial haemorrhage of liver

D. Most affected fish show hyperaemia in brain

[Source: *New Atlas of Fish Diseases*, Kazuo Ogawa]

水霉病（肤霉病）

疾病概述

【概述】 水霉病（肤霉病）是一种能引起水产动物继发性感染的疾病。

【宿主】 主要感染淡水鱼类，如金鱼（*Carassius auratus*）、红鲤（*Cyprinus flammans*）和锦鲤（*Cyprinus carpio*）等观赏鱼类，河蟹、鳖都可患病，对水产动物的种类没有选择性。

【易感阶段】 凡是受伤的水生动物均可被感染，未受伤的不受感染。

【发病水温】 5 ~ 26℃均可发病。有的病原种类甚至在水温30℃时还可生长繁殖。

【地域分布】 世界各地均有分布。

病原

（1）病原是水霉（*Saprolegnia* spp.）和绵霉（*Achlya* spp.）。

（2）均属水霉科（Saprolegniaceae）。

（3）菌丝为管形，没有横隔的多核体。一端为内菌丝，像根样附着在水产动物的损伤处，分枝多而纤细，可深入至损伤、坏死的皮肤及肌肉，具有吸收营养的功能；另一端为外菌丝，较粗壮、分枝较少，可长达3cm，形成肉眼可见的灰白色棉絮状物。

（4）无性生殖时产生动孢子，一般外菌丝的梢端略膨大成棍棒状，同时，内部原生质由下部往这里密集，有性生殖时产生藏卵器和雄器，成为种的重要分类特征。

临床症状和病理学变化

（1）病鱼焦躁不安，与其他固体物发生摩擦，此后鱼体负担过重，游动迟缓，最后瘦弱而死。

（2）河蟹、鳖等患病后食欲减退，行动呆滞，河蟹因无法蜕壳而死亡。

（3）菌丝从伤口侵入后，向外长出外菌丝，似灰白色绵絮状，俗称“生毛”或白毛病。皮肤分泌大量黏液。

（4）在鱼卵孵化过程中，内菌丝侵入卵膜内，卵膜外丛生大量外菌丝，叫“卵丝病”。被寄生的鱼卵，因外菌丝呈放射状，又有“太阳籽”之称。

诊断方法

（1）真菌分离 使用有无菌油菜籽粒的PDA平板，于25℃恒温培养直至油菜籽粒上长满菌丝，将长满菌丝的油菜籽转至装有灭菌过滤河水的6孔板中，分别于15℃和25℃恒温培养。连续2周在倒置显微镜下，观察游动孢子囊、游动孢子的释放及藏卵器和雄器等形态特征可确定水霉种类。

（2）ITS rDNA鉴定 ITS1-F（5′-CTT-GGT-CAT-TTT-AGA-GGA-AGT-AA-3′）和ITS1-R [5′-（TA）TG-GT（CT）-（AGT）（TC）（TC）-TAG-AGG-AAG-TAA-3′]，退火温度为58℃，扩增产物长度为715bp。

预防措施

［鱼体水霉病］

（1）除去池底过多的淤泥，并用200mg/L的生石灰或20mg/L的漂白粉消毒。

（2）加强饲养管理，提高鱼体免疫力，尽量避免鱼体受伤。

（3）亲鱼在人工繁殖时受伤后，可在伤处涂抹10%的高锰酸钾溶液，受伤严重时则须每千克鱼体重肌内或腹腔注射链霉素50 000 ～ 100 000U。

［鱼卵水霉病］

（1）加强亲鱼培育，提高鱼卵受精率，选择晴朗天气进行繁殖。

（2）鱼巢洗净后进行煮沸消毒（棕榈皮做的鱼巢），或用盐、漂白粉等药物消毒（聚草、金鱼藻等做的鱼巢）。

（3）对产卵池及孵化用具进行清洗、消毒。

（4）采用淋水孵化。

（5）鱼巢上黏附的鱼卵不能过多，以免压在下面的鱼卵因得不到足够的氧气而窒息死亡，感染水霉后再进一步危及健康的鱼卵。

诊断方法

（1）全池遍洒食盐及小苏打（碳酸氢钠）合剂（1 ∶ 1），使池水成8mg/L的浓度。

（2）日本鳗鲡在患病早期，可将水温升高到25 ～ 26℃，多数可自愈。

Saprolegniasis (Dermatomycosis)

Disease overview

[Disease Characteristic] Disease causing secondary infection in aquatic animals.

[Susceptible Host] Mainly affect freshwater fish, e. g. ornamental fish such as goldfish (*Carassius auratus*), koi (*Cyprinus carpio*) and carp (*Cyprinus flammans*). River crabs and soft-shell turtles can also be infected. The disease is non-selective to aquatic animal species.

[Susceptible Stage] Infect any traumatized aquatic animals, while not infect those not traumatized.

[Outbreak Water Temperature] Disease outbreaks occur at 5~26 ℃. Some strains can even reproduce at a water temperature of 30℃.

[Geographical Distribution] Widely distribution globally.

Aetiological agents

(1) *Saprolegnia* spp. and *Achlya* spp.

(2) Both of them belong to family Saprolegniaceae.

(3) The fungal hyphae are tubular in shape, with a multinuclear body without septum. One end of the hyphae adheres to the traumatized part of the aquatic animals as root, with numerous tiny branches called internal hyphae that can penetrate deeply into the traumatized and necrotic skin and muscle. These internal hyphae possess the function of nutrient absorption. Those hyphae that protrude the body are external hyphae which are thicker with less branches, up to 3cm in length, forming a greyish white cotton-like floccus.

(4) Form zoospore during asexual reproduction, generally expand as a stick shape at the apex of external hyphae. At the same time, the internal protoplasm condenses at the same location from the lower part. Archegonium and male reproductive organs are formed during the sexual reproductive phase and act as the major classification features of the species.

Clinical signs and pathological changes

(1) Infected fish appears restless and rubs against other solid objects. Afterwards, the fish is overburdened, having slow movement and eventually dies because of emaciation and weakness.

(2) River crabs and soft-shell turtles usually lose appetite and move slowly after being

infected. River crabs die as they cannot undergo the molting process.

(3) The fungal hyphae invade the wound and grow external hyphae exteriorly. The external hyphae has a greyish white cotton-like shape, therefore, it is also called "hair growing" or white hair disease. Large amount of mucus secretions on the skin.

(4) During hatching of fish eggs, the internal hyphae invade the egg membrane while a large amount of external hyphae grows outside of the egg membrane, which is called "egg hyphae disease". The fish eggs being infected are also called 'sun seeds' as the external hyphae show a radiating pattern.

Diagnostic methods

(1) **Fungal isolation** Use a Potato Dextrose Agar (PDA) plate with sterile rapeseed and incubate at 25℃ until the rapeseeds are completely covered with hyphae. Then remove these rapeseeds to a 6-well plate filled with autoclaved filtered river water and incubate at 15℃ and 25℃ separately. Observe continuously for 2 weeks under inverted microscope for zoosporangium, the release of zoospores, the morphological features of archegonium and antheridium, etc., so as to confirm the species of the *Saprolegnia*.

(2) **ITS rDNA sequencing identification** Perform PCR using the primers ITS1-F (5′-CTT-GGT-CAT-TTT-AGA-GGA-AGT-AA-3′) and ITS1-R [5′-(TA)TG-GT(CT)-(AGT)(TC)(TC)-TAG-AGG-AAG-TAA-3′] with annealing temperature at 58℃. The amplicon size is 715bp. Perform sequencing on the amplicon for confirmation.

Preventive measures

[**Saprolegniasis of fish**]

(1) Remove the excessive sludge at the base of the pond and disinfect with 200mg/L quicklime or 20mg/L bleaching powder.

(2) Strengthen aquaculture management, boost fish immunity, and avoid traumatization of fish bodies as far as possible.

(3) If the broodstock is injured during artificial reproduction, the wound can be treated with 10% potassium permanganate solution. In case of severe injuries, inject streptomycin with a dosage of 50,000~100,000U/kg of fish intramuscularly or intraperitoneally for treatment.

[**Saprolegniasis of fish eggs**]

(1) Strengthen cultivation of broodstock, promote the fertilization rate of fish eggs, and select sunny days for breeding.

(2) Disinfect the fish nest (made by palm tree skin) by boiling after washing, or disinfect the fish nest (made by *Myriophyllum spicatum*, *Ceratophyllum*, etc.) by salt, bleaching powder, or other chemicals.

(3) Spawning ponds and tools for hatching should be washed and disinfected.

（4）Implement shower hatching.

（5）The fish eggs adhering on the fish nests cannot be excessive in order to prevent those underneath from suffocation and death due to the lack of oxygen, which would subsequently be infected by *Saprolegnia* and threaten the rest of the healthy eggs.

Treatments

（1）Apply a mixture of sodium chloride and sodium bicarbonate (1 : 1) to the whole pond to form a concentration of 8mg/L in the culture water.

（2）For Japanese eel, the disease is usually self-limiting with full recovery by raising the water temperature to 25~26℃ in early stage of infection.

病鱼细胞及组织病变

A．组织病理切片可见棉絮状菌丝侵入不同组织(HE染色)

B．消化道真菌感染，可见肠道外膜及黏膜均带有大量黑色菌丝(GMS染色)

[源自S．W．Feist及D．Bucke]

Histological lesions of affected fish

A．Fungal hyphae infiltrating the tissues (HE staining)

B．Myocosis in the digestive tract involving the tunica externa and the mucosa. (GMS staining)

[Source：S．W．Feist and D．Bucke]

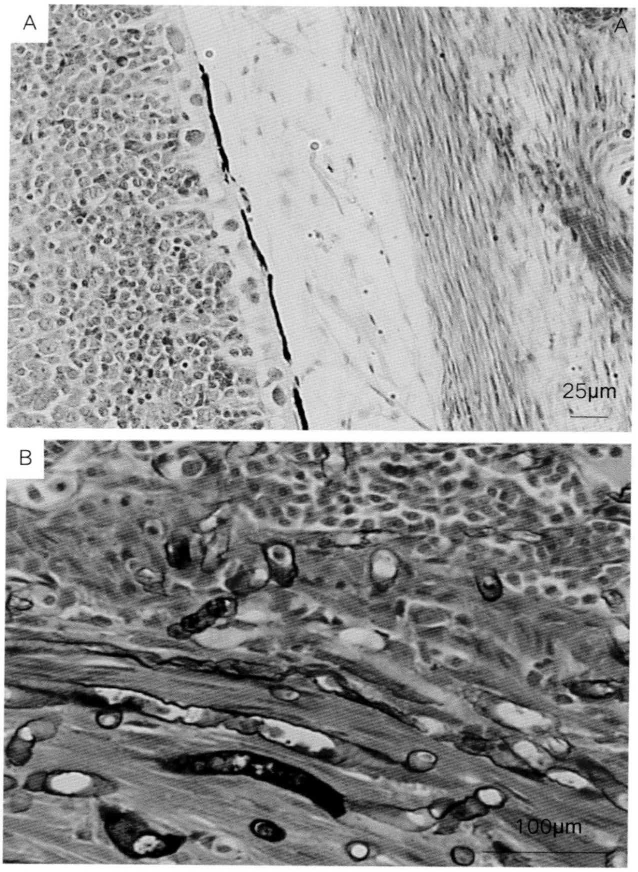

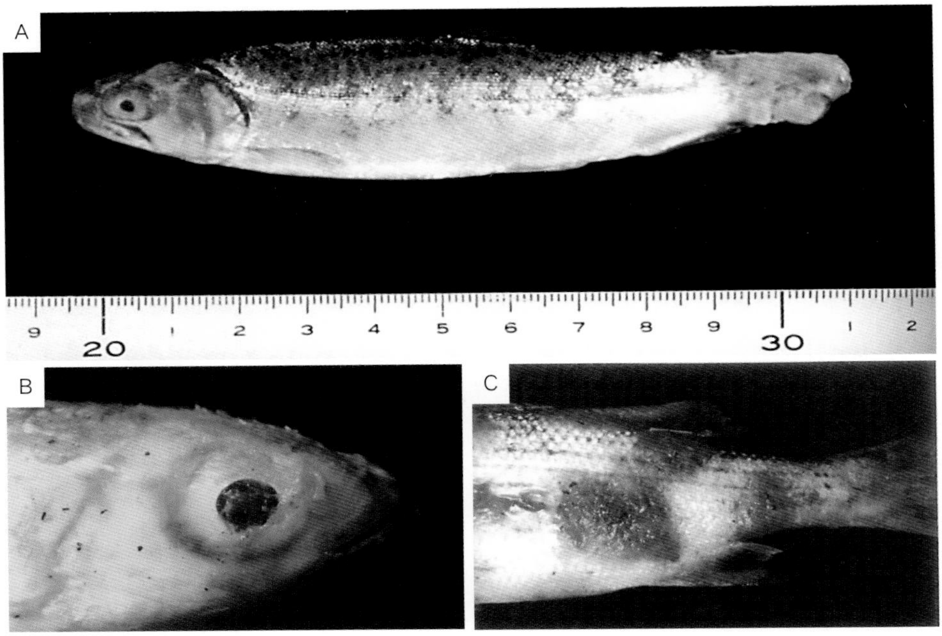

患鱼病变

A．病鱼体表能见到棉絮状菌丝，特别是在头部和尾部

B．乌鱼感染水霉菌，头部附着白色棉絮样物

C．条纹鲈感染水霉菌，体表多处局部区域呈墨绿色，伴随污物附着

［源自《新鱼病图鉴》，小川和夫；《水产动物疾病防治及正确用药手册》，李建霖］

Macroscopic findings of affected fish by Saprolegniasis

A．Fungal hyphae are present on the body surface with higher tendency of infection at the head and tail

B．Fungal hyphae attached to the surface of the head with whitish cotton wadding appearance

C．Dark green patches of *saprolegnia* attached on the skin surface

[Source：*New Atlas of Fish Diseases*，Kazuo Ogawa；*Handbook of Aquatic Animal Disease Control and Medication*，Jianlin Li]

鳃霉病

疾病概述

【概述】 鳃霉病是通过孢子与鳃直接接触而感染的一种疾病。

【宿主】 只对淡水鱼产生危害，主要感染草鱼（*Ctenopharyngodon idellus*）、青鱼（*Mylopharyngodon piceus*）、鳙（*Aristichthys nobilis*）、鲮（*Cirrhinus molitorella*）、银鲴（*Xenocypris argentea* Gunther）、黄颡鱼（*Pelteobagrus fulvidraco*）等。

【易感阶段】 鳃霉对鱼种阶段的鱼敏感，一般鱼苗、鱼种养殖容易发病，病死率可达80%以上，而成鱼不发病或很少发病。

【发病水温】 主要发生在水温超过20℃的高水温季节，严重时呈暴发性死亡。

【地域分布】 流行于世界各地。

病原

（1）病原为鳃霉（*Branchiomyces* spp.）。

（2）属水霉科（Saprolegniaceae）、鳃霉菌属（*Branchimyces*）。

（3）寄生部位为鳃小片或鳃丝血管、软骨，均为组织内寄生。

（4）寄生在草鱼鳃上的鳃霉菌丝较粗直而少弯曲，分枝很少。通常是单枝延长生长，不进入血管和软骨，仅在鳃小片的组织生长，直径为20～25μm，孢子较大，直径为7.4～9.6μm，平均8μm。

（5）寄生在青鱼、鳙、鲮、黄颡鱼鳃上的鳃霉菌丝较细，壁厚，常弯曲成网状，分枝特别多，分枝沿鳃丝血管或穿入软骨生长，纵横交错，充满鳃丝和鳃小片，菌丝直径为6.6～21.6μm，孢子直径为4.8～8.4μm。

临床症状和病理学变化

（1）病鱼失去食欲，呼吸困难，游动缓慢。

（2）鳃上黏液增多，鳃上有出血、瘀血或缺血的斑点，呈现花鳃。

（3）病重时鱼高度贫血，整个鳃呈青灰色。

诊断方法

镜检 用显微镜观察鳃，当发现鳃上有大量鳃霉寄生时，即可做出判断（注意：鳃霉菌丝生长在鳃丝间，特别是穿移鳃霉分枝沿着鳃丝血管或穿入软骨生长时，需要用镊子将鳃丝弯曲和适当“破坏”鳃丝后，在显微镜下才能清晰看见鳃霉菌丝和孢子）。

防治方法

（1）清除池中过多的淤泥，用浓度为450mg/L的生石灰或40mg/L的漂白粉消毒。

（2）严格执行检疫制度，加强饲养管理，注意水质，尤其是疾病流行季节，定期灌注清水，每月全池遍洒1 ~ 2次生石灰（浓度为20mg/L左右）。

（3）掌握投饲量及施肥量，有机肥料必须经发酵后才能放入池中。

（4）排出约2 /3的池水，使用1.5g/m^{3}漂白粉全池泼洒，并于泼洒后12 ~ 24h内加注新水至原水位。如果有必要，可隔1d再重复消毒1次。消毒期间减少或停止投喂并泼洒适量高稳维生素C。待病情基本稳定后，可适当拌料投喂抗菌药物，以防止继发细菌感染。

Branchiomycosis

Disease overview

[Disease Characteristic] A disease transmitted through the direct contact between spores and gills.

[Susceptible Host] Only affect freshwater fish species, mainly grass carp (*Ctenopharyngodon idellus*), black carp (*Mylopharyngodon piceus*), bighead carp (*Aristichthys nobilis*), mud carp (*Cirrhinus molitorella*), blackbelly (*Xenocypris argentea* Gunther), yellow catfish (*Pelteobagrus fulvidraco*), etc.

[Susceptible Stage] Fish at the fingerling stage is particularly sensitive to branchiomycosis. The disease commonly occurs in fry and fingerling aquaculture establishments. Mortality rate of the disease can reach as high as 80% or above, while adult fish does not or seldom show any clinical disease.

[Outbreak Water Temperature] Disease outbreaks mainly occur in warm seasons above 20℃ , which causes massive death in severe cases.

[Geographical Distribution] Widely distributed globally.

Aetiological agents

（1）*Branchiomyces* spp.

（2）Family: Saprolegniaceae. Genus: *Branchimyces*.

（3）Infect gill lamellae or blood vessels/cartilage of gill filaments intracellularly.

(4) *Branchiomyces* spp. infecting grass carp: Hyphae are wide and straight with rare bending and few branches. Usually involve only a single branch which grows by elongation, but not entering blood vessels/cartilage and only grows on gill lamellae. The diameter of hyphae is approximately 20~25μm, the spores are 7.4~9.6μm in diameter with an average of 8μm.

(5) *Branchiomyces* spp. infecting black carp, mud carp and yellow catfish: Hyphae are finer with thick cell wall, bending is common which forms a network structure with a lot of branches. The branches grow along the blood vessels of the gill filaments or into cartilage in a criss-crossed manner, filling up the gill branchial filaments and gill lamellae. The diameter of hyphae is 6.6~21.6μm and the diameter of spores is 4.8~8.4μm.

Clinical signs and pathological changes

(1) Infected fish shows decreased appetite, dyspnoea and slow movement.

(2) Increased mucus secretion of the gills. Haemorrhage, bruises or ischemic spots are also present in the gills, appearing patchy.

(3) In severe case, infected fish suffers from severe anaemia of which the gills appears greenish-grey.

Diagnostic method

Microscopic examination Microscopic examination of gills. If a large number of *Branchiomyces* spp. can be observed on the gills, a diagnosis of branchiomycosis could be made. (Note: The hyphae of *Branchiomyces* spp. grow in-between the gill filaments, especially *Branchiomyces demigrans*. Its branches grow along the blood vessels of the gill filaments or into the cartilage, where a forceps would be required to bend and appropriately "damage" the gill filaments to expose the hyphae and spores of *Branchiomyces* spp. clearly under microscope.)

Preventive measures

(1) Remove excessive sludge from the pond and disinfect with 450mg/L quicklime or 40mg/L bleaching powder.

(2) Apply stringent quarantine measures. Strengthen aquaculture management. Pay close attention to the water quality, especially during the peak season of the disease. Refill clean water to the pond regularly and apply quicklime (with the concentration of around 20mg/L) to the whole pond once or twice monthly.

(3) Control and monitor the feeding and fertilization situation. Organic fertilizers must be fermented before being applied.

(4) Drain out around 2/3 of the culture water and apply 1.5g/m^3 bleaching powder to the whole pond. Then add new water to the original water level within 12~24 hours. The disinfection

can be repeated once after one day if needed. During the disinfection period, reduce or stop feeding and supply an appropriate amount of highly stable vitamin C, which can help in reducing the mortality. After the disease situation is stabilized, antimicrobial agents can be appropriately applied in the feed for prevention of secondary bacterial infection.

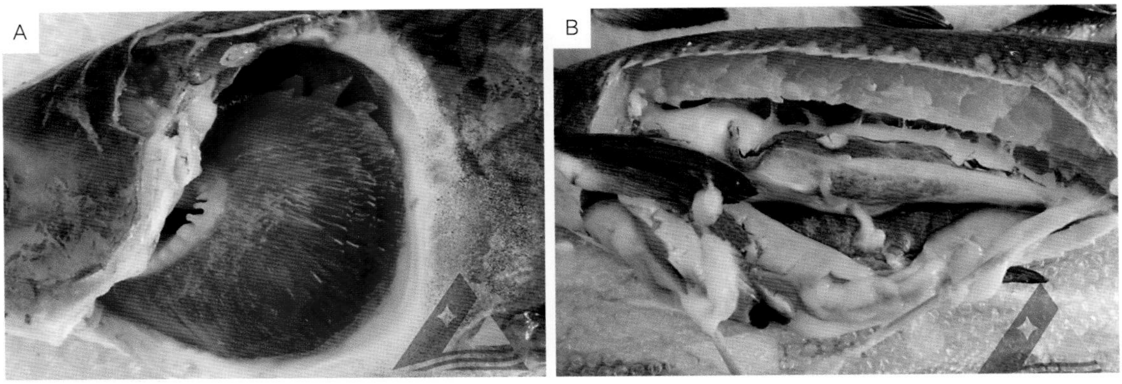

患鱼病变

A. 鳃丝水肿，充血、贫血、坏死，呈花鳃症状　B. 肝脏发黄

[源自利洋]

Macroscopic findings of affected fish

A. Gill filament oedema, with patchy hyperaemia, anaemia and necrosis　B. Liver appears pale yellow in color

[Source：LIYANG AQUATIC]

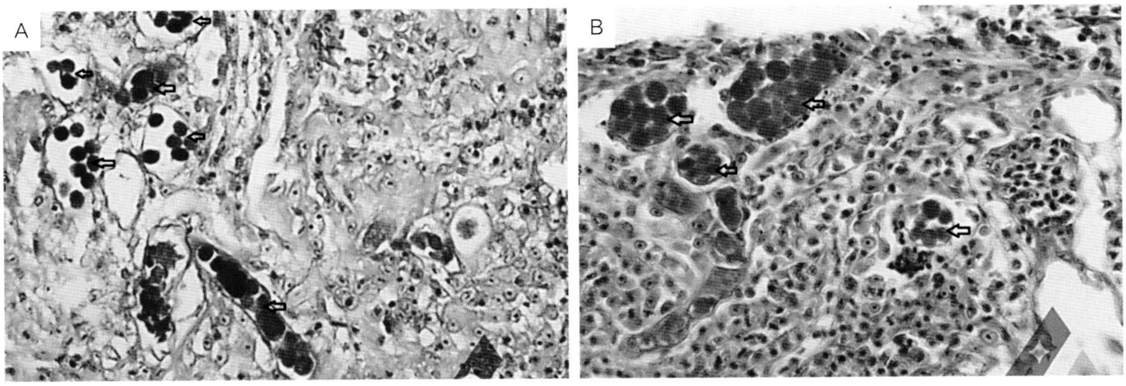

病鱼细胞及组织病变

A、B. 鳃丝上的菌丝和孢子（箭头）（HE染色）

[源自利洋]

Histological lesions of affected fish

A&B. Fungal hyphae and spores (arrow) present in the gill (HE staining)

[Source：LIYANG AQUATIC]

流行性溃疡综合征

疾病概述

【概述】 流行性溃疡综合征也叫红点病（RSD）或霉菌性肉芽肿病（MG）或溃疡性霉菌病（UM），是一种野生和人工养殖淡水鱼及河口鱼的流行病。

【易感宿主】 首次在香鱼（*Plecoglossus altivelis*）中发现，可自然感染黄鳍鲷（*Acanthopagrus australis*）、黑鲷（*Acanthopagrus berda*）、昌达鲈（*Ambassis agassiz*）、黑色大头鱼（*Ameiurus melas*）、横纹羊鯻（*Amniataba percoides*）、攀鲈（*Anabas testudineus*）、海鲇（*Arius* sp.）、麦克利栉鳞鳎（*Aseraggodes macleayanus*）、直鳍鲃（*Barbus peludinosus*）、耀尾鲃（*Barbus poechii*）、塔马拉鲃（*Barbus thamalakanensis*）、无纹鲃（*Barbus unitaeniatus*）、银鲈（*Bidyanus bidyanus*）、大西洋鲱（*Brevoortia tyrannus*）、侧身非洲脂鲤（*Brycinus lateralis*）、卡特拉鲃（*Catla catla*）、纹鳢（*Channa striatus*）、印度鲮（*Cirrhinus mrigala*）、尖齿胡鲇（*Clarias gariepinus*）、钝齿胡鲇（*Clarias ngamensis*）、蟾胡子鲇（*Clarius batrachus*）、矮鲈（*Colisa lalia*）、条鳍鱼（*Esomus* sp.）、黄鳝（*Fluta alba*）、舌天竺鲷（*Glossamia aprion*）、舌虾虎鱼（*Glossogobius giuris*）、非洲梭子鱼（*Hepsetus odoe*）、非洲虎鱼（*Hydrocynus vittatus*）、斑点叉尾鮰（*Ictalurus punctatus*）、苗圃鱼（*Kurtus gulliveri*）、红眼鲮（*Labeo cylindricus*）、新月野鲮（*Labeo lunatus*）、南亚野鲮（*Labeo rohita*）、尖吻鲈（*Lates calcarifer*）、单色匀鯻（*Leiopotherapon unicolor*）、翻车鱼（*Lepomis macrochirus*）、紫红笛鲷（*Lutjanus argentimaculatus*）、大鳞异吻象鼻鱼（*Marcusenius macrolepidotus*）、绿锦鱼（*Melanotaenia splendida*）、尖齿小鲑脂鲤（*Micralestes acutidens*）、乌鱼（*Mugil cephalus*）、鲻科鱼（Mugilidae）、北澳海鰶（*Nematalosa erebi*）、三斑罗非鱼（*Oreochromis andersoni*）、绿头罗非鱼（*Oreochromis machrochir*）、大丝足鲈（*Osphronemus goramy*）、线纹尖塘鳢（*Oxyeleotris lineolatus*）、石头鱼（*Oxyeleotris marmoratus*）、岩头长颌鱼（*Petrocephalus catostoma*）、牛鳅（*Platycephalus fuscus*）、大菱鲆（*Psettodes* sp.）、爪哇鲤（*Puntius gonionotus*）、斑尾小鲃（*Puntius sophore*）、露鲃（*Rohtee* sp.）、彩虹鲷（*Sargochromis carlottae*）、绿鲷（*Sargochromis codringtonii*）、粉红鲷（*Sargochromis giardia*）、金钱鱼（*Scatophagus argus*）、银锡伯鲇（*Schilbe intermedius*）、非洲黄油鲇（*Schilbe mystus*）、乔氏硬骨舌鱼（*Scleropages jardini*）、条纹钱蝶鱼（*Selenotoca multifasciata*）、窄头丽鱼（*Serranochromis angusticeps*）、粗壮丽鱼（*Serranochromis robustus*）、鳍（*Sillago ciliata*）、井鲇和玻璃鲇（Siluridae）、克氏柱颌针鱼（*Strongylura kreffti*）、鯻（*Therapon* sp.）、伦氏非鲫（*Tilapia rendalli*）、斯氏非鲫（*Tilapia sparrmanii*）、查达射水鱼（*Toxotes chatareus*）、洛氏射水鱼（*Toxotes lorentzi*）、粗鳞毛足鲈（*Trichogaster pectoralis*）、线足鲈（*Trichogaster trichopterus*）。尼罗罗非鱼（*Oreochromis niloticus*）、遮目鱼（*Chanos chanos*）、鲤（*Cyprinus carpio*）等对这种病有抗性。

【易感阶段】 快速生长阶段和个体成熟阶段的鱼易感，现未发现鱼苗和幼鱼发生。

【发病水温】 在低温季节发病持续时间长。暴雨过后、低温或温度介于18 ～ 22℃时极易暴发。

【地域分布】 主要流行于日本、美国、澳大利亚东部、巴布亚新几内亚以及东南亚、南亚、非洲南部、欧洲等国家和地区。

【疾病地位】 世界动物卫生组织（OIE）将其列入水生动物疫病名录。

病原

（1）病原为丝囊霉菌（*Aphanomyces invadans*）。

（2）属细囊霉科（Leptolegniaceae）、丝囊霉属（*Aphanomyces*）。

（3）具有无隔的真菌样菌丝体结构，有2种典型的游走孢子形式。初级游走孢子由在孢子囊内发育的圆形细胞组成，于孢子囊的顶部释放，并在此形成孢子群，其很快转换成次级游走孢子。肾形次级游走孢子具有横向双鞭毛细胞，可在水中自由游动。

（4）无性繁殖，在体外6℃下可以生长，最适生长温度为20 ～ 30℃，在体外37℃下不能生长。

（5）丝囊霉基因组全长为71.4Mb，共15 416个基因，编码20 816个蛋白。

临床症状和病理学变化

（1）患病鱼早期不吃食，鱼体发黑，漂浮在水面上，有时会不停地游动。

（2）在体表、头、鳃盖和尾部可见红斑。在后期会出现较大的红色或灰色的浅部溃疡，并常伴有棕色的坏死。

（3）躯干和背部发生大块的损伤。

（4）对于特别敏感的鱼，损伤会逐渐扩展加深，以至身体较深的部位，或者造成头盖骨软组织和硬组织的坏死，使活鱼的脑部暴露出来。

（5）引起强烈的炎症反应，并在长入肌肉的菌丝周围形成肉芽肿。刮开损伤部位后通常可以看到霉菌、细菌或寄生虫的二次感染。除了肌肉组织和表皮外，其他组织如肝脏、肾脏、脾脏、胰腺、鳔、肠、腹膜、性腺等也会观察到有菌丝生长。

诊断方法

（1）组织病理学鉴定 组织病理切片中呈现霉菌性肉芽肿或内部肌肉组织中分离到丝囊霉。

（2）显微镜观察 当损伤由慢性温和皮炎发展为局部严重扩展的坏死性肉芽肿皮炎，并使肌肉严重絮片状退化时，可见丝囊霉无隔的菌丝样结构。

（3）分离鉴定

①中等大小的病鱼，取2mm左右的病灶组织，使用加有100U/mL的青霉素和100μg/mL的噁喹酸Czapek Dox培养基，25℃培养。将菌丝在新鲜的Czapek Dox培养基上重复转接，

直至得到纯培养物。

②小于20cm的病鱼，取2～4mm病灶组织，使用加有100U/mL青霉素和100μg/mL噁喹酸GP（葡萄糖/蛋白胨）的平板，25℃下培养并在12h内镜检，将出现的菌丝重复转移到含1.2%琼脂的加有100U/mL青霉素和10μg/mL链霉素的新鲜GP平板，直到培养物纯净为止，并于10℃在GP培养基上保存，或者培养不超过7d。

（4）PCR　有3对引物可供选择：

①Ainvad-2F（5′-TCA-TTG-TGA-GTG-AAA-CGG-TG-3′）　和Ainvad-ITSR1（5′-GGC-TAA-GGT-TTC-AGT-AGT-ATG-TAG-3′），退火温度为56℃，扩增产物长度为234bp。

②ITS11（5′-GCC-GAA-GTT-TCG-CAA-GAA-AC-3′）和ITS23（5′-CGT-ATA-GAC-ACA-AGC-ACA-CCA-3′），退火温度为65℃，扩增产物长度为550bp。

③BO73（5′-CTT-GTG-CTG-AGC-TCA-CAC-TC-3′）和BO639（5′-ACA-CCA-GAT-TAC-ACT-ATC-TC-3′），退火温度为58℃，扩增产物长度为564bp。

扩增产物经测序后判定。

防治方法

（1）大型水体中暴发，病情几乎不受控制。

（2）在小水体和封闭水体里暴发，通过清除病鱼、用生石灰消毒池水，从而改善水质。

（3）在养鱼场中使用氯化钠或盐类及农用石灰。

（4）腹腔注射含维生素C 300g/kg、维生素B 150g/kg及微量的维生素B_1、维生素B_2、维生素B_6和维生素B_{12}的免疫增强剂（Salar-bec），可增强血清抑制游走孢子萌芽及生长的能力。

Epizootic ulcerative syndrome, EUS

Disease overview

[Disease Characteristic] Seasonal epidemic disease affecting wild and captive freshwater and brackish water fish species, also called red spot disease (RSD) or mycotic granulomatous disease (MG) or ulcerative mycosis (UM).

[Susceptible Host] First reported in sweetfish (*Plecoglossus altivelis*). Naturally affect yellowfin bream (*Acanthopagrus australis*), black bream (*Acanthopagrus berda*), Agassiz’s perchlet (*Ambassis agassiz*), black bullhead (*Ameiurus melas*), barred grunter (*Amniataba percoides*), climbing perch (*Anabas testudineus*), fork-tailed seafishes (*Arius* sp.), narrow banded sole (*Aseraggodes macleayanus*), straightfin barb (*Barbus peludinosus*), dashtail barb (*Barbus poechii*), thalamakane barb (*Barbus thamalakanensis*), slender barb (*Barbus unitaeniatus*), silver perch (*Bidyanus bidyanus*), Atlantic menhaden (*Brevoortia tyrannus*), striped robber (*Brycinus lateralis*), catla (*Catla catla*), snakehead murrel (*Channa striatus*), mrigal carp (*Cirrhinus mrigala*), African sharptooth catfish (*Clarias gariepinus*), blunt-toothed African catfish (*Clarias ngamensis*), walking catfish (*Clarius batrachus*), walking catfish (*Colisa lalia*), flying barb (*Esomus* sp.), swamp eel (*Fluta alba*), cardinaldishes (*Glossamia aprion*), tank goby (*Glossogobius giuris*), African pike characin (*Hepsetus odoe*), African tigerfish (*Hydrocynus vittatus*), channel catfish (*Ictalurus punctatus*), nurseryfish (*Kurtus gulliveri*), redeye labeo (*Labeo cylindricus*), Upper Zambezi labeo (*Labeo lunatus*), roho labeo (*Labeo rohita*), barramudi (*Lates calcarifer*), spangled perch (*Leiopotherapon unicolor*), bluegill (*Lepomis macrochirus*), mangrove red snapper (*Lutjanus argentimaculatus*), bulldog (*Marcusenius macrolepidotus*), eastern rainbowfish (*Melanotaenia splendida*), sharptooth tetra (*Micralestes acutidens*), flathead grey mullet (*Mugil cephalus*), Mullet (Mugilidae), bony bream (*Nematalosa erebi*), three-spotted tilapia (*Oreochromis andersoni*), longfin tilapia (*Oreochromis machrochir*), giant gourami (*Osphronemus goramy*), sleepy cod (*Oxyeleotris lineolatus*), marble goby (*Oxyeleotris marmoratus*), Churchill (*Petrocephalus catostoma*), black flathead (*Platycephalus fuscus*), spiny turbot (*Psettodes* sp.), java barb (*Puntius gonionotus*), pool barb (*Puntius sophore*), *Rohtee* sp., rainbow happy (*Sargochromis carlottae*), green happy (*Sargochromis codringtonii*), pink bream (*Sargochromis giardia*), spotted scat (*Scatophagus argus*), silver butter catfish (*Schilbe intermedius*), African butter catfish (*Schilbe mystus*), Australian bonytongue (*Scleropages jardini*), spotbanded scat (*Selenotoca multifasciata*), thinface cichlid (*Serranochromis angusticeps*), yellow-belly bream (*Serranochromis robustus*), sand whiting (*Sillago ciliate*), wel and glass catfishes (Siluridae), long tom (*Strongylura kreffti*), *Therapon* sp., redbreast tilapia (*Tilapia rendalli*), banded tilapia (*Tilapia sparrmanii*), seven-spot archerfish (*Toxotes chatareus*), primitive archerfish (*Toxotes lorentzi*), snakeskin gourami (*Trichogaster pectoralis*), three-spot gourami (*Trichogaster trichopterus*). Nile tilapia (*Oreochromis niloticus*), milkfish (*Chanos chanos*) and common carp (*Cyprinus carpio*) are resistant to this disease.

[Susceptible Stage] Fish at rapid growing stage and maturing stage are susceptible to the disease. No disease report on fry and fingerlings.

[Outbreak Water Temperature] Disease persists in low temperature seasons for a long period. Outbreaks easily occur after heavy rain, low temperature environment or when water temperature at 18~22℃ .

[Geographical Distribution] Mainly epidemic in Japan, USA, Eastern Australia, Papua New Guinea, Southeast Asia, South Asia, Southern Africa, Europe, etc.

[Disease Status] OIE-listed Aquatic Animal Disease.

Aetiological agent

(1) *Aphanomycas invadans*.

(2) Family: Leptolegniaceae. Genus: *Aphanomyces*.

(3) Possess a fungal-like mycelium structure without septum. Two classical types of

zoospores. Primary zoospores are formed by the round cells that develop within the sporangium and released from the top of the sporangium, forming a spore group there. They soon transform into secondary zoospores. Kidney-shaped secondary zoospores have double laterally flagellated cells, freely movable in water.

(4) Asexual reproduction. *In vitro* growth occurs at 6℃ with optimal growth at 20~30℃ . No *in vitro* growth at 37℃ .

(5) The genome size of *Aphanomyces* is 71.4Mb consisting of 15,416 genes, encoding 20,816 proteins.

Clinical signs and pathological changes

(1) At early stage of the disease, affected fish are anorexic, show darkening of the body trunk, float on water surface and sometimes swim restlessly.

(2) Reddish patchy lesions are present on body surface, head, gill operculum and tail. At later stage, larger red or grey superficial ulcers can be seen, and usually accompanied with brownish necrosis.

(3) Large areas of ulcers on the body trunk and the back.

(4) For exceptionally sensitive fish, the ulcerative lesions will spread and deepen progressively to deeper parts of the body, or causes necrosis of the soft tissues and hard tissues of the cranium, exposing the brain of the live fish.

(5) Severe inflammatory response forming granulomas surrounding the *Aphanomyces*-invaded muscles. Dissecting these lesions usually reveals secondary fungal, bacterial or parasitic infections. Apart from muscles and skin, fungal hyphae are also present in other internal organs such as liver, kidney, pancreas, swimming bladder, intestines, mesenteries and gonads.

Diagnostic methods

(1) Histopathology examination Observe for granulomatous lesions accompanied with fungal hyphae resembling morphology of *Aphanomyces* on histology sections, or *Aphanomyces* can be isolated from internal muscle tissues.

(2) Microscopic examination Fungal hyphae resembling morphology of *Aphanomyces* can be observed when the lesions progress from mild chronic dermatitis to severe locally extensive granulomatous dermatitis causing marked distorting of the muscular architecture.

(3) Fungal isolation

① Moderate-sized fish: Collect around 2mm of sample from relevant lesions and place it to Czapek Dox agar with 100U/mL penicillin and 100μg/mL oxolinic acid, and incubate at 25℃ . Transfer cultured hyphae to a fresh Czapek Dox agar plate repeatedly until a pure culture is obtained.

② Fish less than 20cm: Collect around 2~4mm of sample from relevant lesions, place the sample on a GP (glucose/peptone) plate with 100U/mL penicillin and 100μg/mL oxolinic acid and incubate at 25℃. Examine with microscope within 12 hours. Repeatedly transfer the appearing hyphae to GP (glucose/peptone) plates containing 1.2% agar, with 100 U/mL penicillin and 100μg/mL oxolinic acid until the fungal culture is pure. Store in GP agar at 10℃ or continuous passage with intervals of no longer than 7 days.

(4) **Polymerase chain reaction (PCR)** There are 3 pairs of primers available.

① Ainvad-2F (5′-TCA-TTG-TGA-GTG-AAA-CGG-TG-3′) and Ainvad-ITSR1 (5′-GGC-TAA-GGT-TTC-AGT-AGT-ATG-TAG-3′) with annealing temperature at 56℃. The amplicon size is 234bp.

② ITS11 (5′-GCC-GAA-GTT-TCG-CAA-GAA-AC-3′) and ITS23 (5′-CGT-ATA-GAC-ACA-AGC-ACA-CCA-3′) with annealing temperature at 65℃. The amplicon size is 550bp.

③ BO73 (5′-CTT-GTG-CTG-AGC-TCA-CAC-TC-3′) and BO639 (5′-ACA-CCA-GAT-TAC-ACT-ATC-TC-3′) with annealing temperature at 58℃. The amplicon size is 564bp.

Perform sequencing on the amplicons for confirmation.

Preventive measures

(1) If the disease outbreaks occur in large compartment, the disease is almost uncontrollable.

(2) If the disease outbreaks in small compartment with closed water system, the mortality of the disease can be reduced effectively by eliminating infected fish, disinfection of water with quicklime and improving water quality.

(3) Application of sodium chloride or other salts and agricultural lime in aquaculture establishments is a safe and effective treatment, which can also prevent the transmission of EUS.

(4) Intraperitoneal injections of immunostimulant "Salar-bec" (containing vitamin C 300g/kg, vitamin B 150g/kg and trace amount of vitamin B_1, vitamin B_2, vitamin B_6 and vitamin B_{12}) can enhance immunity of fish to inhibit zoospores from budding and growing.

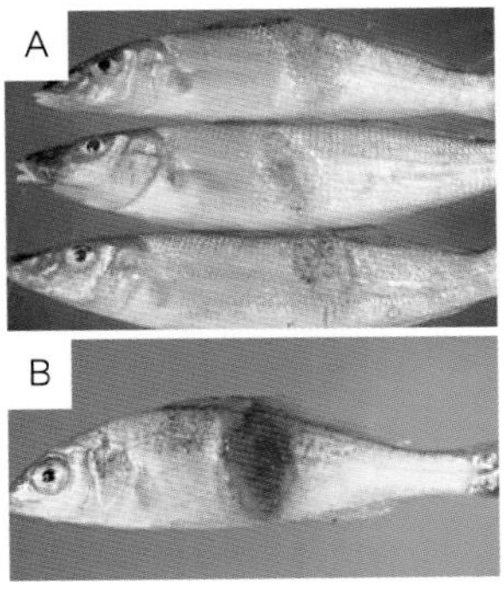

患鱼病变

A. 感染EUS的鳕，可见病变从皮肤发红（上）发展到深度溃疡（下）

B. 幼银鲈患EUS可见红色溃疡

[源自New South Wales Department of Primary Industries]

Macroscopic findings of affected fish. Fish affected by EUS are showing various degrees of skin ulceration on the body trunk

A. Cod fish B. Juvenile of silver perch

[Source: New South Wales Department of Primary Industries]

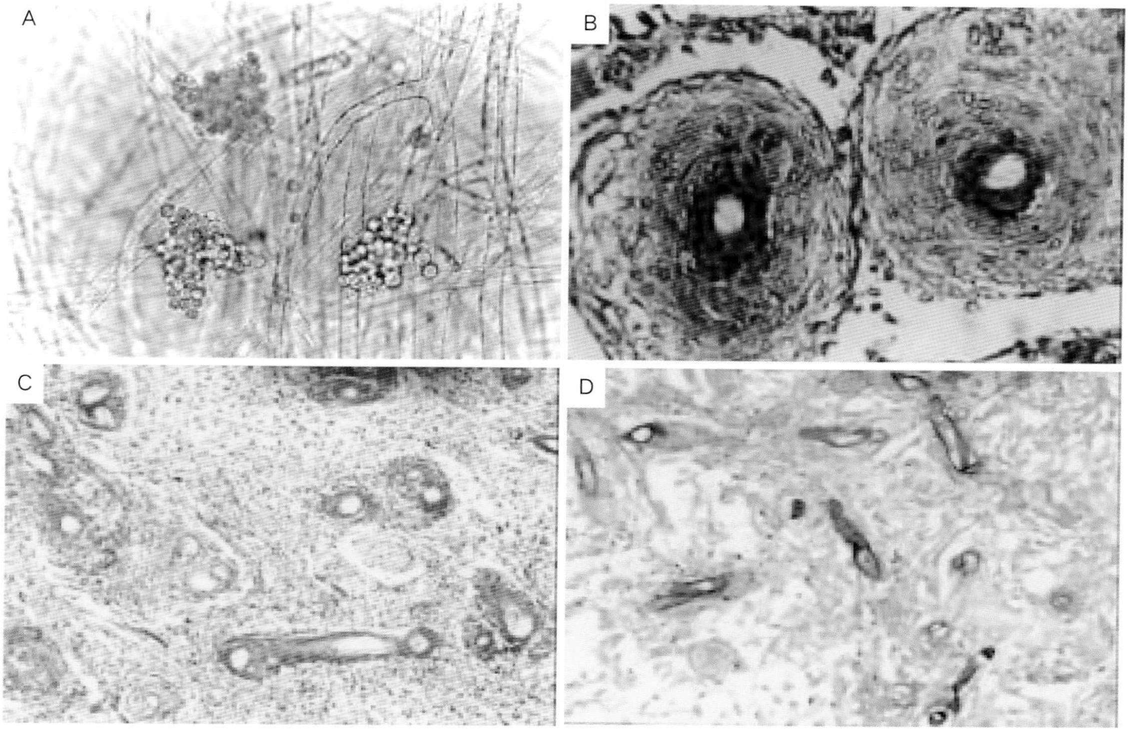

病鱼细胞及组织病变

A．EUS真菌的涂片制备，可见典型的丝囊霉菌

B&C．组织病理观察可见坏死、肉芽肿性皮炎和肌膜炎（HE染色）D．真菌菌丝涂片(Grocott染色)

[源自ENACA]

Cytology findings and histological lesions of affected fish

A．Direct smear of lesion showing typical spores of *Aphanomycetes*　B&C．Necrosis and granulomatous inflammation with refractile organisms in the tissue sections (HE staining)　D．Fungal hyphae are present in the lesions (Grocott staining)

[Source：ENACA]

竖 鳞 病

疾病概述

【概述】 竖鳞病又称鳞立病、松鳞病、松球病等，是一种鱼类常见病。

【易感宿主】 主要感染金鱼（*Carassius auratus*）、鲫（*Carassius auratus*）、鲤（*Cyprinus carpio*）、草鱼（*Ctenopharyngodon idellus*）、鲢（*Hypophthalmichthys molitrix*）、鳙（*Aristichthys nobilis*）以及各种热带鱼等。

【易感阶段】 从大规格鱼种到亲鱼均可受害。

【发病水温】 水温在17～22℃时易发生，有时在越冬后期也有发生。死亡率一般在50%以上，病情严重的鱼池，甚至100%死亡。鲤亲鱼的死亡率可高达85%。

【地域分布】 分布于世界各地，流行于静水养鱼池和高密度养殖条件下，流水养鱼池中较少发生。

病原

（1）一般认为，病原为水型点状假单胞菌（*Pseudomonas punctate* f. *ascitae*）。但有人认为，病原为气单胞菌或类似这一类的细菌；也有人认为是一种循环系统疾病，由淋巴回流障碍而引起。

（2）水型点状甲单胞菌隶属于假单胞菌科（Pseudomonadaceae）、假单胞菌属（*Pseudomonas*）。

（3）短杆状，近圆形，单个排列，无芽孢，具有运动能力，革兰氏阴性菌。

（4）细菌菌落呈圆形，培养24h后中等大小，略带黄灰色，在光线下透明、呈培养基的颜色

临床症状和病理学变化

（1）病鱼离群独游，游动缓慢，无力，严重时呼吸困难，对外界刺激失去反应，身体失去平衡，身体倒转，腹部向上，浮于水面。

（2）疾病早期鱼体发黑，体表粗糙，鱼体前部的鳞片竖立，向外张开像松球，而鳞片基部的鳞囊水肿，它的内部积聚着半透明的渗出液。严重时全身鳞片竖起，鳞囊内积有含血的渗出液，用手轻压鳞片，渗出液就从鳞片下喷射出来，鳞片也随之脱落。

（3）伴有鳍基、皮肤轻微充血，眼球突出，腹部膨大，腹腔内积有腹水，鳍条间有半透明液体，顺着与鳍条平行的方向用力压之，液体即喷射出来。

（4）病鱼贫血，鳃、肝、脾、肾的颜色均变淡。

诊断方法

（1）镜检 根据临床症状，同时镜检鳞囊内的渗出液，见有大量革兰氏阴性短杆菌即可做出诊断。

（2）细菌分离 使用普通琼脂培养基，28℃下培养12 ~ 24h可见菌落呈圆形，24h培养后形成中等大小、光滑、略黄而稍灰白、边缘隆起不透明的菌落。在TSA上菌落圆形光滑、边缘整齐、浅灰白色，未见水溶性褐色色素产生。在麦康凯培养基上，在初始划线位置上仅微弱生长薄细菌或无色菌落。

（3）生化鉴定 可根据《伯杰氏手册》或API细菌鉴定系统、全自动微生物鉴定系统进行生化鉴定，主要生化特性为革兰氏阴性，氧化酶阳性，有运动性，可分解葡萄糖、蔗糖、甘露糖、阿拉伯糖，不分解乳糖、尿素，V-P试验阳性，靛基质试验阴性，硝酸盐还原阴性，七叶苷试验阳性。

（4）16S rDNA鉴定 27F（5′-AGA-GTT-TGA-TCC-TGG-CTC-AG-3′）和1 492R（5′-GGT-TAC-CTT-GTT-ACG-ACT-T-3′），退火温度为55℃，扩增长度为1 500bp。

扩增产物经测序后判定。

预防方法

（1）在捕捞、运输、放养时，勿使鱼体受伤。

（2）发病初期冲注新水，可使病情停止蔓延。

（3）以浓度为5mg/L的硫酸铜、2mg/L的硫酸亚铁和10mg/L的漂白粉混合液浸洗鱼体5 ~ 10min。

（4）将病原菌制成灭活菌苗，通过注射菌苗，可使鱼类获得对该病较高的免疫力。

（5）每50kg水加入捣烂的大蒜250g，浸洗病鱼数次。

（6）用3%的食盐水浸洗病鱼10 ~ 15min，或用2%食盐、3%小苏打混合液浸洗10min。

治疗方法

（1）每千克鱼体重每天用磺胺二甲氧嘧啶100 ~ 200mg，连用3 ~ 5d。

（2）轻轻压破鳞囊的水肿泡，勿使鳞片脱落，用10%的温盐水擦洗，再涂抹碘酊。同时，肌内注射磺胺嘧啶钠2mL。

（3）每千克鱼体重每天用复方新诺明（复方磺胺甲基异噁唑）100mg，拌饲投喂。第二天开始药量减半，连用5d。

（4）用氟苯尼考拌料投喂并全池泼洒，同时，配合敌百虫杀灭鱼虱。

Lepmorthosis

Disease overview

[Disease Characteristic] A common disease in fish, also called "Scale erecting disease" "Pine scale disease" "Pine cone disease" .

[Susceptible Host] Mainly affect goldfish (*Carassius auratus*), common carp (*Cyprinus carpio*), grass carp (*Ctenopharyngodon idellus*), silver carp (*Hypophthalmichthys molitrix*), bighead carp (*Aristichthys nobilis*) and different species of tropical fishes.

[Susceptible Stage] Age stages from fingerlings to broodstocks are all susceptible to this disease.

[Outbreak Water Temperature] Disease outbreaks easily occur when water temperature at 17~22℃ , sometimes also occur in late winter. Mortality rate is usually higher than 50%, while aquaculture farms with severe disease may have a mortality rate of 100%. The mortality rate of broodstock of common carp can reach 85%.

[Geographical Distribution] Widely distributed globally. Endemic in stillwater aquaculture establishment and farms with high cultivation density, seldom occurs in aquaculture establishments with running water.

Aetiological agents

(1) Believed to be *Pseudomonas punctata* f. *ascitae*. It is considered that the aetiological agent is *Aeromonas* spp. or a similar bacterium. There are also some says that it is a kind of circulatory systemic disease, caused by a disorder in lymphatic backflow.

(2) Pseudomonas *punctata* f. *ascitae* belongs to genus, *Pseudomonas* under family, Pseudomonadaceae.

(3) Gram-negative short-rod-shaped bacterium, almost circular, arranged singly, motile, non-spore forming.

(4) Bacterial colonies are circular in shape, moderately sized after incubated for 24 hours, slightly yellowish gray, transparent under light revealing the color of the media.

Clinical signs and pathological changes

(1) Segregated, slow movement, and weakness. In severe infection, the fish shows dyspnea,

loses response to external stimuli and balance of the body, turns upside down, and floats on the water surface.

(2) At early stage of the disease, the fish body darkened with roughened body surface. The scales at the anterior part of the fish body stand upright, radiating outwards like a pine cone. The scale sacs at the base become oedematous with accumulation of translucent exudate. In severe case, the scales of the whole body stand upright, with bloody exudate accumulated inside the scale sacs. If the scales are being compressed gently, the exudate will eject out accompanied with detachment of scales.

(3) Mild haemorrhage of the base of the fins and the skin, exophthalmos, abdominal distension with ascites accumulation, translucent secretion between fin rays which ejects by compression parallel to the fin rays.

(4) Anaemia with paling of the gill, liver, spleen and kidney.

Diagnostic methods

(1) **Microscopic diagnosis** Based on clinical signs, and microscopic examination of the exudate within the scale follicles. Diagnosis could be made when a large number of Gram-negative short-rod bacteria are present.

(2) **Bacterial isolation** Inoculate on general culture agar plate and incubate at 28 ℃ for 12~24 hours, forming round colonies. After 24 hour inoculation, colonies are medium sized. Smooth, light yellow to greyish white in color, with irregular borders. On TSA agar, the bacteria form round, smooth, irregularly-edges, greyish white colonies without water-soluble brownish pigments. On MacConkey agar, only weakly growth thin bacterial lawn and colorless colonies are present at the initial streaking position.

(3) **Biochemical identification** With reference to *Bergey's Manual* or API identification system or automatic microbial indentification system to conduct microbiology biochemical identification. Major biochemical characteristics are: gram-negative; oxidase positive; motile; glucose-, sucrose-, mannose- and arabinose-fermenting, but not fermenting galactose and urease; negative for V-P reaction; indole test negative; nitrate reduction test negative; and esculine test positive.

(4) **16S rDNA sequencing identification** Use the primers 27F (5′-AGA-GTT-TGA-TCC-TGG-CTC-AG-3′) and 1,492R (5′-GGT-TAC-CTT-GTT-ACG-ACT-T-3′) to perform PCR amplification with annealing temperature at 55℃ . The amplicon size is 1,500bp. Perform sequencing on the amplicon for confirmation.

Preventive measures

(1) During capturing, transporting and stocking, avoid traumatizing the fish.

(2) Refill new water at the early stage of disease to prevent spreading of disease.

(3) Immerse affected fish in a mixed solution of 5mg/L copper sulfate, 2mg/L ferrous sulfate and 10mg/L bleach for 5~10 min.

(4) Vaccination of inactivated vaccine produced by the isolated pathogen. By vaccination, a higher immunity against this disease can be achieved.

(5) For every 50kg of water, add 250g grinded garlic and immerse affected fish to treat for several times.

(6) Immerse affected fish in 3% sodium chloride solution for 10~15 min or mixed solution of 2% sodium chloride and 3% baking soda for 10 min.

Treatment

(1) Sulfadimethoxine: 100~200 mg/kg of fish per day for consecutive 3~5 days.

(2) Gently compress and rupture the oedematous cysts of the scale follicles but not detaching the scales. Then wash with 10% warm saline, followed by topical application of iodine tincture. Simultaneously inject 2mL of sulfadiazine sodium intramuscularly.

(3) Sulfamethoxazole: 100mg/kg of fish in feed per day, reduce dose by half from the second day onwards and treat for consecutive 5 days.

(4) Mix florfenicol in feed and apply to the whole pond, simultaneously use trichlorfon to kill fish lices.

患鱼病变

全身鳞片竖立

[源自R.Herbert及汪开毓]

Macroscopic findings of affected fish

The scales along the body trunk of the affected fish stand upright, with a pine cone-like appearance

[Source: R.Herbert and Kaiyu Wang]

分枝杆菌病

疾病概述

【概述】 分枝杆菌病是一种慢性疾病。

【宿主】 主要感染鲤（*Cyprinus carpio*）、金鳟（*Oncorhynchus mykiss*）、胡子鲇（*Clarias fuscus*）等淡水养殖的鱼以及大菱鲆（*Scophthalmus maximus*）、牙鲆（*Paralichthys olivaceus*）、欧洲舌齿鲈（*Dicentrarchus labrax*）、金点篮子鱼（*Siganus punctatus*）、纹条蝴蝶鱼（*Chaetodon falcula*）、尖吻重牙鲷（*Diplodus puntazzo*）、金头鲷（*Sparus aurata*）、波纹石斑鱼（*Epinephelus ongus*）、眼斑拟石首鱼（*Sciaenops ocellatus*）、鲻（*Mugil cephalus*）、五条鰤（*Seriola quinqueradiata*）、裸胸鳝（*Gymnothorax* sp.）等海水养殖的鱼。

【易感阶段】 各年龄阶段均易感。

【发病水温】 水温在30 ~ 32℃时最易感。

【地域分布】 流行于世界各地，包括西班牙、葡萄牙、美国、马来西亚、南非、英国、意大利、斯洛文尼亚等。

病原

（1）病原为海分枝杆菌（*Mycobacterium marinum*）、龟分枝杆菌（*M. chelonae*）和偶发分枝杆菌（*M. fortuitum*）。

（2）均属分枝杆菌科（Mycobacteriaceae）、分枝杆菌属（*Mycobacterium*）。

（3）龟分枝杆菌和偶发分枝杆菌属于速生型分枝杆菌（RGM），培养时间一般少于7d；海分枝杆菌属于缓慢生长型分枝杆菌（SGM），培养时间通常大于7d，需要培养2 ~ 4 周才形成清晰可见的菌落。

（4）杆状，大小不一，为（0.2 ~ 0.6）μm×（1 ~ 10）μm，无鞭毛、无芽孢、无荚膜，革兰氏阳性菌，为好氧型。

（5）生长缓慢，一般培养基上不生长，在有光处培养时菌落呈黄色或橘黄色，最适生长温度为30 ~ 32℃，37℃以上生长缓慢或不生长。

（6）海分枝杆菌属腐生型、非典型、缓慢生长型分枝杆菌，是条件致病菌，培养温度为25 ~ 30℃，在勒-詹二氏培养基（LJ培养基）上产生平滑到粗糙的菌落；生长在黑暗处的菌落无色素。幼龄菌落生长在光照下或短暂曝光时，变为鲜黄色。

（7）龟分枝杆菌属速生型分枝杆菌、环境腐生菌。呈多形态，大小为（0.2 ~ 0.5）μm×（1 ~ 6）μm，培养温度为22 ~ 40℃，42℃以上不生长。

（8）偶发分枝杆菌属速生型分枝杆菌，菌体大小为（0.2 ~ 0.6）μm×（1 ~ 3）μm，培养温度为25 ~ 37℃。许多菌株在40℃和22℃时生长，但是在45℃和17℃时生长受到抑制，

45℃以上不生长。在凝缩的卵培养基中接种2 ～ 4d，生长状态良好。

临床症状和病理学变化

（1）最初症状是皮肤溃疡、红肿，形成小结节，随着病情的发展，在内脏中也形成许多灰白色或淡黄褐色的小结节，肝脏、肾脏、脾脏等器官形成小的坏死病灶。

（2）雌鱼的卵巢受到侵害时，鱼卵发生退行性变性。

诊断方法

（1）细菌分离 采集感染鱼的肝脏、脾脏或血液样本，将细菌接种到勒-詹二氏培养基和Middlebrook 7H10+OADC琼脂培养基分离，30℃培养30d。挑取单菌落用 Middlebrook 7H10琼脂培养基纯化。

（2）镜检 取内脏中的小结节做涂片，进行抗酸染色后，镜检发现长杆形的抗酸菌，结合临床症状，可确诊。

（3）生化鉴定 可根据《伯杰氏手册》或API细菌鉴定系统、全自动微生物鉴定系统进行生化鉴定，主要生化特性为耐热触酶阳性，硝酸盐还原阴性，吐温80水解试验阳性，尿素酶试验阳性，芳香硫酸酯酶阳性，亚硝酸盐还原试验阳性，甘露醇利用阴性，草酸盐利用阴性，阿拉伯糖产酸试验阴性，木糖产酸阴性，柠檬酸盐利用阴性等。

（4）巢式PCR

第一步引物为：T39（5′-GCG-AAC-GGG-TGA-GTA-ACA-CG-3′）和T13（5′-TGC-ACA-CAG-GCC-ACA-AGG-GA-3′），退火温度为50℃，扩增产物长度为936pb。

第二步引物为：preT43（5′-AAT-GGG-CGC-AAG-CAA-GCC-TGA-TG-3′）和T531（5′-ACC-GCT-ACA-CCA-GGA-AT-3′），退火温度为50℃，扩增产物长度为300 ～ 312bp。

扩增产物测序后判定。

（5）高效液相色谱法（HPLC） 商品化的HPLC分析分枝菌酸系统可鉴定分枝杆菌。

防治方法

（1）加强鱼类养殖环境中的分枝杆菌检测，做好水体中分枝杆菌的动态分布研究。

（2）选择有效的含氯消毒剂，降低养殖环境中分枝杆菌的数量。

（3）不用患分枝杆菌病的鱼作饲料，或先将鱼煮熟后再喂。

（4）每天每千克鱼体重用强力霉素或复方新诺明（复方磺胺甲基异噁唑）100mg左右，连续投喂10 ～ 14d。

Mycobacteriosis

Disease overview

[Disease Characteristic] Chronic disease.

[Susceptible Host] Mainly affects freshwater fish species such as common carp (*Cyprinus carpio*), rainbow trout (*Oncorhynchus mykiss*) and catfish (*Clarias fuscus*); as well as marine fish species such as turbot (*Scophthalmus maximus*), olive flounder (*Paralichthys olivaceus*), European bass (*Dicentrarchus labrax*), goldspotted spinefoot (*Siganus punctatus*), blackwedged butterflyfish (*Chaetodon falcula*), sheephead bream (*Diplodus puntazzo*), gilt-head bream (*Sparus aurata*), white-streaked grouper (*Epinephelus ongus*), red drum (*Sciaenops ocellatus*), flathead grey mullet (*Mugil cephalus*), Japanese amberjack (*Seriola quinqueradiata*) and moray eels (*Gymnothorax* sp.).

[Susceptible Stage] All age groups are susceptible.

[Outbreak Water Temperature] Fish are most susceptible to the disease at a water temperature of 30~32℃ .

[Geographical Distribution] Distributed globally, including Spain, Portugal, USA, Malaysia, South Africa, England, Italy and Slovenia, etc.

Aetiological agents

(1) *Mycobacterium marinum*, *M. chelonae* and *M. fortuitum*.

(2) All of them belongs to genus *Mycobacterium* under family Mycobacteriaceae.

(3) *M. chelonae* and *M. fortuitum* are rapidly growing mycobacteria (RGM) that the incubation time usually less than 7 days, while *M. marinum* is a slowing growing mycobacterium (SGM) that the incubation time usually larger than 7 days and may take 2~4 weeks for visible colonies.

(4) Gram-positive rod-shaped aerobic bacteria, without flagella, non-spore forming, and without capsule. Different in sizes, ranging from (0.2~0.6)μm × (1~10)μm.

(5) Slow growth and no growth on usual media. The colonies show a yellow or orange-yellow color when being incubated in places under light. The optimal growing temperature is 30~32℃ , slow or no growth when the temperature is above 37℃ .

(6) *M. marinum* is a saprophytic and atypical SGM, and is an opportunistic pathogen. The incubation temperature is 25~30℃ . It grows smooth to rough colonies on Lowenstein Jensen (LJ) medium which are non-pigmented when growing in darkness. If the young colonies are exposed to

light, the colonies will appear yellow.

(7) *M. chelonae* is an environmental saprophytic RGM with polymorphism of (0.2~0.5)μm × (1~6)μm in size. The incubation temperature is 22~40℃ , no growth above 42℃ .

(8) *M. fortuitum* is an RGM of (0.2~0.4)μm × (1~3)μm in size. The incubation temperature is 25~37℃ , many strains grow at 40℃ and 22℃ but with inhibited growth at 45℃ and 17℃ , no growth above 45℃ . Grow well in LJ medium with incubation time of 2~4 days.

Clinical signs and pathological changes

(1) The initial clinical signs are skin ulcers and reddening, forming small nodules. As the disease progresses, many grayish white or yellowish brown small nodules are also formed in the internal organs, sometimes small necrotic lesions are also formed in the liver, kidney, spleen, etc.

(2) When the ovaries of female fish are invaded, the fish eggs will have degenerative changes.

Diagnostic methods

(1) **Bacterial isolation** Collect liver, spleen or blood samples from the infected fish aseptically to isolate the pathogenic bacteria. Inoculate the samples to LJ medium and Middlebrook 7H10 + OADC agar medium for isolation and incubate at 30℃ for 30 days. Pick a single colony and perform secondary culture on Middlebrook 7H10 + OADC agar medium for purification.

(2) **Microscopic examination** Take the small nodules from internal organs to prepare direct smears and perform acid-fast staining on the smears. If long-rod-shaped acid-fast bacteria are present with consistent clinical signs, a confirmatory diagnosis could be made.

(3) **Biochemical identification** With reference to *Bergey's Manual* or API identification system or automatic microbial indentification system to conduct microbiology biochemical identification. Major biochemical characteristics are: catalase heat stable test positive, nitrate reduction test negative, tween 80 hydrolysis test positive, urease test positive, arylsulfatase test positive, nitrite reduction test positive, mannitol fermentation negative, oxalate utilization test negative, acid production from arabinose fermenting test negative, acid production from xylose test negative and citrate test negative, etc.

(4) **Nested Polymerase Chain Reaction (PCR)**

① First step primers: T39 (5′-GCG-AAC-GGG-TGA-GTA-ACA-CG-3′) and T13 (5′-TGC-ACA-CAG-GCC-ACA-AGG-GA-3′) with annealing temperature at 50℃. The amplicon size is 936bp.

② Second step primers: preT43 (5′-AAT-GGG-CGC-AAG-CAA-GCC-TGA-TG-3′) and T531 (5′-ACC-GCT-ACA-CCA-GGA-AT-3′) with annealing temperature at 50℃ . The amplicon size is 300~312bp.

Perform sequencing on the amplicon for confirmation.

(5) **High performance liquid chromatography (HPLC)** Identify mycobacteria by commercialized HPLC analytical system.

Preventive measures

(1) Strengthen mycobacterial screening in aquaculture environment and study the dynamic distribution of mycobacteria in water bodies.

(2) Select effective chlorinated disinfectants to minimize the number of mycobacteria in aquaculture environment.

(3) Do not use *Mycobacterium*-infected fish as feed, or thoroughly cook the bait fish before feeding.

(4) Doxycycline or sulfamethoxazole: Apply at a dosage of around 100 mg/kg of fish per day and treat orally for consecutive 10~14 days.

病鱼内脏

肝脏、肾脏、脾脏、心脏、消化道、鳃、肌肉有许多大大小小的白色结节

[源自《新鱼病图鉴》，小川和夫]

Macroscopic findings of affected fish

Numerous whitish nodules are present in the internal organs of affected fish

[Source: *New Atlas of Fish Diseases*, Kazwo Ogawa]

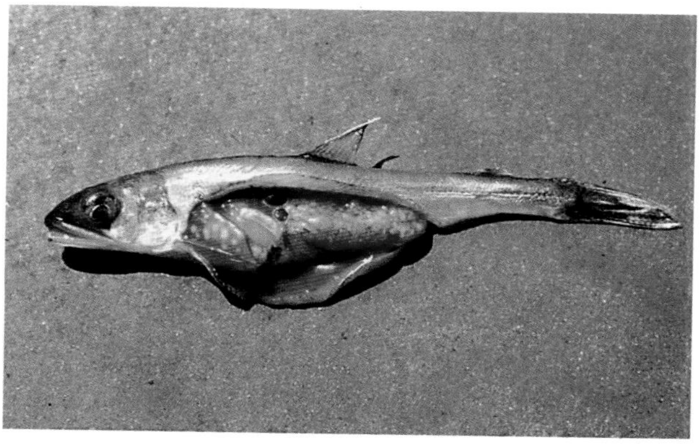

培养病原菌形成菌落

将病灶用4%氢氧化钠处理，用LJ培养基培养约4周后，可长出灰白色菌落

[源自《新鱼病图鉴》，Kazuo Ogawa]

Pathogenic bacteria grew into colonies by culturing

Greyish white colonies cultured on Lowenstein-Jensen medium from samples treated with 4% sodium hydroxide and incubate for about 4 weeks

[Source: *New Atlas of Fish Diseases*, Kazuo Ogawa]

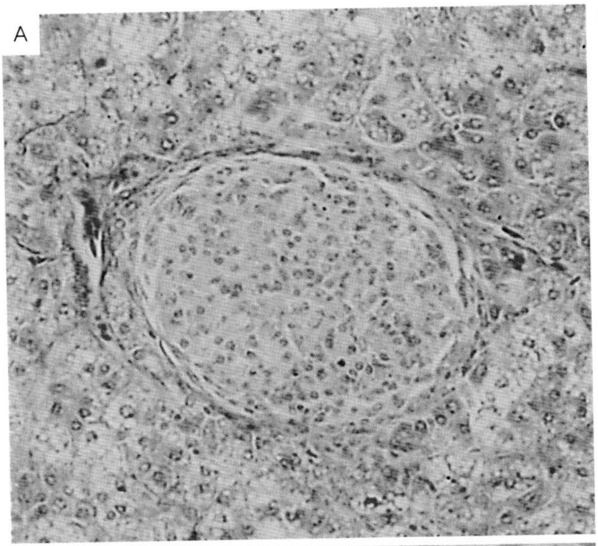

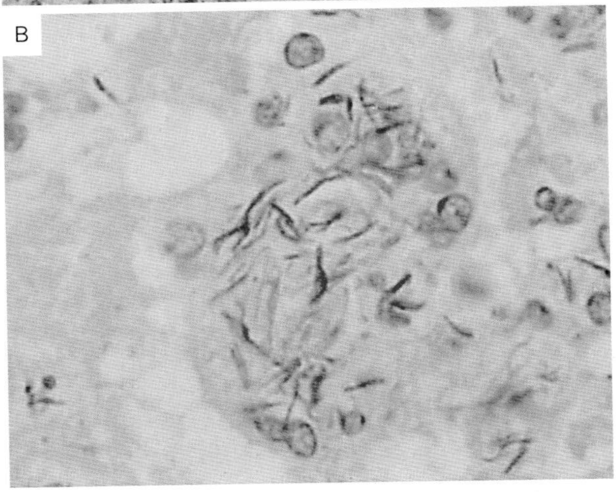

病鱼组织病变

A．对白色结节进行病理组织学观察时，可见到肉芽肿病变（HE染色）

B．在肉芽肿内见有抗酸性的长杆菌（ZN染色）

[源自《新鱼病图鉴》，小川和夫]

Histological lesions of affected fish

A．Granuloma are present in multiple internal organs (HE staining)

B．Numerous acid-fast bacilli are present in the granuloma (ZN staining)

[Source：*New Atlas of Fish Diseases*，Kazuo Ogawa]

诺卡氏菌病

疾病概述

【概述】 诺卡氏菌病是一种危害严重的细菌病。

【宿主】 可感染淡水鱼和海水鱼，如花鲈（*Lateolabrax japonicus*）、卵形鲳鲹（*Trachinotus ovatus*）、青斑鮨（*Hypoplectrus indigo*）、鮸（*Miichthys miiuy*）、红鱼（*Lutjanus erythopterus*）、美国红鱼（*Sciaenops ocellatus*）、尖吻鲈（*Lates calcarifer*）、泥鲾（*Siganusfus cescens*）、大口黑鲈（*Micropterus salmoides*）、黑鱼（*Channa argus*）、大黄鱼（*Pseudosciaena crocea*）、比目鱼（*Heterosomata* sp.）、五条鰤（*Seriola quinqueradiata*）、高体鰤（*Seriola dumerili*）等。

【易感阶段】 主要危害6个月至2龄鱼，可导致慢性死亡，死亡率10%～30%。

【发病水温】 水温在15～32℃时都可流行，水温25～28℃时为流行高峰。

【地域分布】 首次发现于1963年美洲阿根廷的淡水热带鱼，现广泛流行于世界各地。

病原

（1）病原为诺卡氏菌（*Nocardia* spp.），有鰤诺卡氏菌（*N. seriolae*）、星状诺卡氏菌（*N. asteroides*）、粗形诺卡氏菌（*N. crassostreae*）、杀鲑诺卡氏菌（*N. salmoncide*）。

（2）属类诺卡氏菌科（Nocardiaceae）、类诺卡氏菌属（*Nocardia*）。

（3）菌体呈长或段杆状，或细长分枝状，常断裂成杆状至球状体，基丝发达，呈分枝状，气丝较少，具膨大或棒状末端。直径0.2～1.0μm，长2.0～5.0μm。丝状体长10～50μm，不生孢子，无运动能力，革兰氏阳性菌，专性需氧菌。

（4）生长缓慢，生长温度范围为12～32℃，最适温度为25～28℃。形成白色或淡黄色沙粒状菌落，粗糙易碎，边缘不整齐，偶尔在表面形成皱褶。

临床症状和病理学变化

（1）病鱼大体分为躯干结节型和鳃结节型两类，两种类型的病鱼心脏、脾脏、肾脏、鳔等处都有结节。

（2）躯干结节型的病鱼在躯干部的皮下脂肪组织和肌肉发生脓疡，在外观上则膨大突出成为许多大小不一、形状不规则的结节，剖开疖疮后流出白色或略带红色的脓汁。病理组织切片可出现病灶的周围多数有成层的纤维芽细胞。

（3）鳃结节型的病鱼在鳃丝基部形成乳白色的大型结节，鳃明显褪色。

诊断方法

（1）镜检 从病鱼结节处取少量脓汁制成涂片，进行革兰氏染色。镜检发现有阳性的丝状菌，结合临床症状基本可以确诊。

（2）ITS rRNA鉴定

① 128F（5′-CAG-TCC-GAT-ATC-GCG-GTG-AA-3′）和128R（5′-GAC-CTA-CTC-GAC-CAA-GTG-GC-3′），退火温度为59℃，扩增长度为128bp。

② Noc-F（5′-CAC-CTA-CGA-AAA-TCC-CAT-TTG-GT-3′）和Noc-R（5′- CAT-CGG-ATT-GGA-TTC-AAG-GAC-CTT-GA-3′），退火温度为60℃，扩增产物长度为156bp。

扩增产物经测序后判定。

防治方法

（1）投饵勿过量，避免养殖水体富营养化或残饵堆积。

（2）每天每千克鱼体重用土霉素70 ～ 80mg，制成药饵，连续投喂5 ～ 7d。

（3）在口服药饵的同时，用漂白粉等消毒剂全池泼洒，视病情用1 ～ 2次，可以提高防治效果。

Nocardiosis

Disease overview

[Disease Characteristic] Severe bacterial disease of fish.

[Susceptible Host] Can affect both freshwater and marine fish species such as sea bass (*Lateolabrax japonicas*), pompano (*Trachinotus ovatus*), indigo hamlet (*Hypoplectrus indigo*), Chinese drum (*Miichthys miiuy*), crimson snapper (*Lutjanus erythropterus*), red drum (*Sciaenops ocellatus*), barramundi (*Lates calcarifer*), pinspotted spinefoot (*Siganus fuscescens*), largemouth bass (*Micropterus salmoides*), northern snakehead (*Channa argus*), large yellow croaker (*Pseudosciaena crocea*), flatfish (*Heterosomata* sp.), Japanese amberjack (*Seriola quinqueradiata*) and greater amberjack (*Seriola dumerili*).

[Susceptible Stage] Mainly affect fish of 6-month to 2-year old, causing chronic mortality of 10%~30%.

[Outbreak Water Temperature] Disease outbreaks occur at water temperature of 15~32℃ with peak at 25~28℃ .

[Geographical Distribution] First reported in freshwater tropical fish in Argentina in 1963, now distributed globally.

Aetiological agents

(1) *Nocardia* spp., including *N. seriolae*, *N. asteroides*, *N. crassostreae* and *N. salmoncide*.

(2) Family: Nocardiaceae. Genus: *Nocardia*.

(3) Gram-positive obligate aerobic bacteria, elongated or short-rod in shape, sometimes branching with fragmented branches appearing rod or coccobacilli in shape, possessing well-developed branching substrate mycelium and smaller number of aerial mycelium with bulged ends. The bacterial is approximately 0.2~1.0μm in diameter, 2.0~5.0μm in length with mycelium of 10~50μm long, no-spore forming and non-motile.

(4) Slow-growing at temperature ranges from 12~32℃, with optimal temperature at 25~28℃. It forms white or light yellow sand-like colonies, which are rough and fragile with irregular edges, sometimes folds on the surface.

Clinical signs and pathological changes

(1) The presentation of the disease can be sub-categorised into two types: body trunk furunculosis and gill furunculosis. Both form nodular lesions in the heart, spleen, kidney, swimming bladder, etc., in affected fish.

(2) Body trunk furunculosis: Abscessation at the subcutaneous tissue and muscle of the body trunk, externally swells up and protrudes as many non-uniform sized, non-uniform shaped nodules, or called furuncles. White or slightly reddish purulent discharge are present inside these furuncles if dissected. Histologically, there are usually layers of fibroblast cells surrounding the lesions.

(3) Gill furunculosis: Milky white large nodules are formed at the base of the gill filaments. Marked paling of the gill.

Diagnostic methods

(1) Microscopic examination Take a small amount of purulent discharge from the nodules of affected fish to prepare direct smear and perform Gram staining. If microscopic examination reveals Gram-positive filamentous bacteria, in combination of suggestive clinical signs, a confirmatory diagnosis could be made.

(2) ITS rRNA polymerase chain reaction (PCR) There are 2 pairs of primers available.

① 128F (5′-CAG-TCC-GAT-ATC-GCG-GTG-AA-3′) and 128R (5′-GAC-CTA-CTC-GAC-CAA-GTG-GC-3′) with annealing temperature at 59℃. The amplicon size is 128bp.

② Noc-F (5′-CAC-CTA-CGA-AAA-TCC-CAT-TTG-GT-3′) and Noc-R (5′- CAT-CGG-ATT-GGA-TTC-AAG-GAC-CTT-GA-3′) with annealing temperature at 60℃. The amplicon size is

156bp.

Perform sequencing on the amplicon for confirmation.

Preventive measures

(1) Avoid overfeeding to prevent the culture water from eutrophication or accumulation of residual feed.

(2) Oxytetracycline: 70~80 mg/kg of fish per day, prepare as medicated feed and treat consecutively for 5~7 days.

(3) Accompany with oral treatment, apply bleaching powder or other disinfectant to the whole pond for disinfection once to twice to enhance the preventive effects depending on the disease situation.

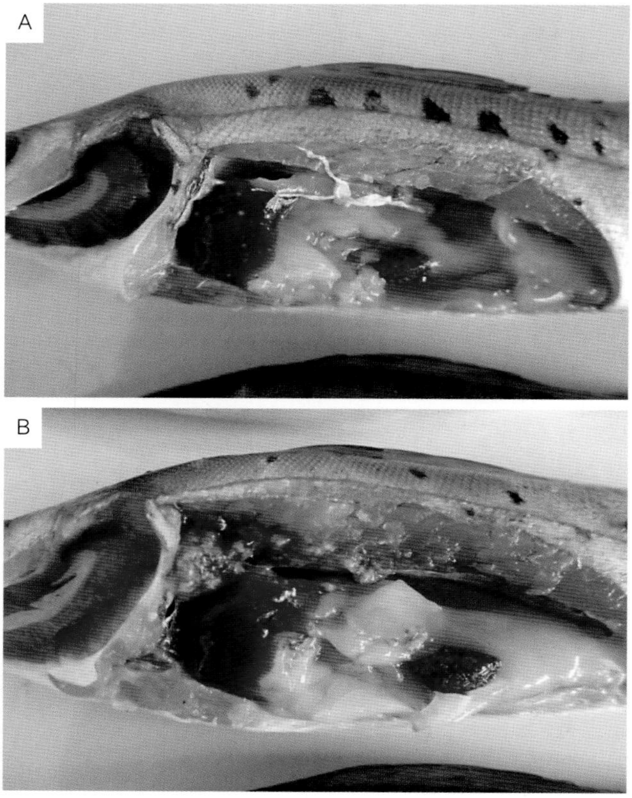

患诺卡氏菌病的海鲈

A. 烂鳃，肝脏、脾脏上有白色结节状增生

B. 肝脏、脾脏、肾脏、脂肪上有白色结节状增生，鳃色浅

[源自唐绍林]

Macroscopic findings of affected sea bass

A. Severe gill necrosis. Multiple pinpoint whitish nodules present in liver and spleen

B. Pale gill. Numerous pinpoint whitish nodules present in liver, spleen, kidney and visceral adipose tissues

[Source: Shaolin Tang]

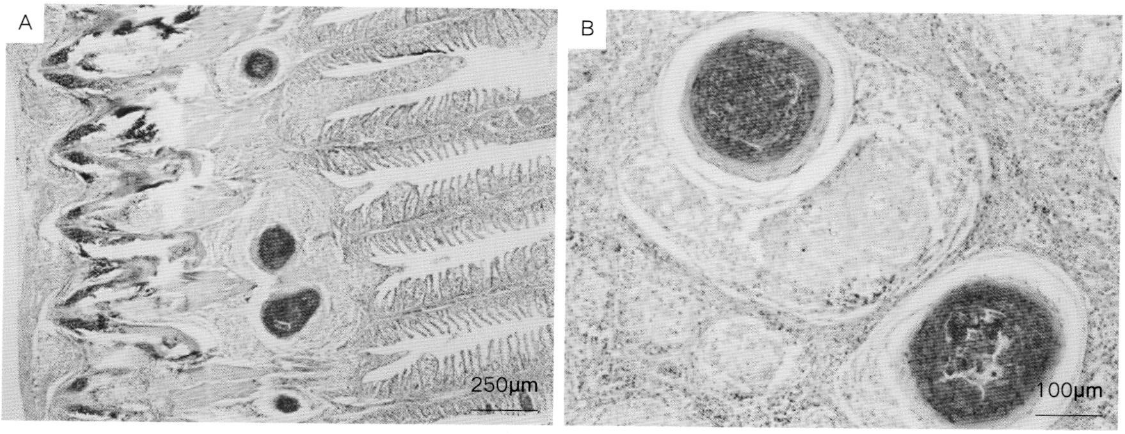

感染卡诺氏菌病的丽鱼细胞及组织病变（HE染色）

A、B．组织连接处有肉芽肿病变

[源自安迪·荷里曼]

Histological lesions of affected fish (HE staining)

A&B．Granulomatous lesions in the connective tissues of the gills

[Source：Andy Holliman]

爱德华氏菌病

疾病概述

【概述】 爱德华氏菌病是鱼类一种严重的传染病。迟缓爱德华氏菌主要引起鲇气肿性腐败病（EPDC），鮰爱德华氏菌主要引起斑点叉尾鮰肠型败血症（ESC）。这两种爱德华氏菌都可引起黄颡鱼红头病。

【宿主】

（1）迟缓爱德华氏菌 主要感染多种淡水鱼和海水鱼，如日本鳗鲡（*Anguilla japonica*）、牙鲆（*Paralichthys olivaceus*）、尖吻鲈（*Lates calcarifer*）、斑点叉尾鮰（*Ictalurus punctatus*）、大口黑鲈（*Micropterus salmoides*）、胭脂鱼（*Mugil cephalus*）、红海鲷（*Evynnis japonica*）、尼罗罗非鱼（*Tilapia nilotica*）、大鳞大麻哈鱼（*Oncorhynchus tshawytscha*）、真鲷（*Pagrus major*）、黄体鰤（*Seriola lalandi*）、虹鳟（*Oncorhynchus mykiss*）、鲤（*Cyprinus carpio*）、欧洲舌齿鲈（*Dicentrarchus labrax*）、大菱鲆（*Scophthalmus maximus*）、蟾胡子鲇（*Clarias batrachus*）、美洲红点鲑（*Salvelinus fontinalis*）、卡特拉鲃（*Catla catla*）、南亚野鲮（*Labeo rohita*）、欧洲鳗鲡（*Anguilla anguilla*）、远东鲇（*Silurus asotus*）等。此外，该菌还可感染无脊椎动物、两栖动物、爬行动物、鸟类和哺乳动物。

（2）鮰爱德华氏菌 主要感染鲇形目鱼类，最易感染斑点叉尾鮰（*Ictalurus punctatus*）；还可感染鲇形目的其他种类，如犀目鮰（*Ameiurus catus*）、蟾胡子鲇（*Clarias batrachus*）和云斑鮰（*Ictalurus nebulosus*）；也可感染非鲇形目鱼类，如日本鳗鲡（*Anguilla japonica*）和观赏性鱼类中的斑马鱼（*Barchydanio rerio*）、玫瑰无须鲃（*Puntius conchonius*）、蓝色弓背鱼（*Notopterus notopterus*）等。人工接种下，可感染大鳞大麻哈鱼（*Oncorhynchus tshawytscha*）和虹鳟（*Oncorhynchus mykiss*）。

【易感阶段】 迟缓爱德华氏菌主要危害体重450g以上的鱼，鮰爱德华氏菌可感染各个年龄阶段的鱼。

【发病水温】 感染可发生于全年，在水温15℃时就能发生，高峰期多出现在水温25～30℃时。EPDC在水温超过30℃时少暴发；ESC的发病温度为20～30℃，或者更低温。黄颡鱼红头病发病水温为18～30℃，在20～28℃暴发流行，环境条件恶化时在30℃以上可发病。

【地域分布】 呈世界性分布，主要分布在澳大利亚、印度、马来群岛、以色列、日本、巴拿马、美国以及其他热带和亚热带地区；对日本和中国台湾养鳗业的危害尤其大。

病原

（1）病原为迟缓爱德华氏菌（*Edwardsiella tarda*）和鲇爱德华氏菌（*E. ictaluri*）。其中，

迟缓爱德华氏菌是一种重要的人畜共患病的病原菌。

（2）属哈夫尼亚菌科（Enterobacteriaceae）、爱德华氏菌属（*Edwardsiella*）。

（3）短杆状，大小多在（0.5～1)μm×(1～3)μm，不形成芽孢，无荚膜，有周身鞭毛，能运动。但从真鲷和鲱中分离的病原无运动性，革兰氏阴性菌，兼性厌氧，是一种条件致病菌。

（4）生长温度范围为15～42℃，迟缓爱德华氏菌、鮰爱德华氏菌最适生长温度分别为37℃、25～30℃。

（5）迟缓爱德华氏菌具有菌体（O）抗原和鞭毛（H）抗原。根据O抗原，将迟缓爱德华氏菌分为A、B、C、D 4个血清型，A型致病性较强。目前，一般认为鮰爱德华氏菌的血清型只有1个；但也有报道认为，根据来源不同（鲇形目鱼类与非鲇形目鱼类），存在至少2种不同的血清型和DNA质粒类型。

临床症状和病理学变化

（1）感染迟缓爱德华氏菌的病鱼，发病初期仅能在病鱼体后外侧观察到直径为3～5mm的皮肤损伤。严重时，胸腹和尾柄部的肌肉可出现溃疡，且病灶迅速扩大，并形成空洞，形成外观隆起、肿胀，内部充满气体的气肿。患处常见褪色，若切开病灶，会有恶臭的气体释放出，坏死的组织可填满空洞1/3的体积。

（2）鮰爱德华氏菌感染，可分为慢性型和急性型感染：

①慢性型：经神经系统感染，形成肉芽肿性炎症，引起慢性脑膜炎，病鱼精神沉郁，食欲减退，浮于水面，有时以头朝上、尾朝下的姿势悬停于水中，并伴有打转或狂游。病灶周围出现褪色现象，两眼间出现1条纵向的溃疡灶，最后溃疡灶加深，暴露出头骨，在头部形成1个开放性的溃疡。

②急性型：经消化道侵入血液，病鱼腹部膨大，体表、肌肉可见细小的充血、出血、积液。流出的腹水不易凝固，肝脏水肿，质脆，有出血点和灰白色的坏死斑点，脾、肾肿大并出血，胃膨大，肠道扩张，充血、发炎，肠腔内充满气体和淡黄色水样液体，黏膜水肿、充血。

（3）患红头病的黄颡鱼离群独游、反应迟钝，食欲退减，腹部膨大，鳍条基部、下颌、鳃盖、腹部、肛门及生殖孔充血、出血。解剖腹腔，内含有大量含血或清亮的液体，肝肿大，有出血点或出血斑，肾脏充血、肿大，脾脏肿大呈紫黑色。肠道扩张、充血发炎，肠腔内充满气体和淡黄色水样液体。

（4）组织病理学上表现为间质性肾炎、化脓性肝炎和脾脏化脓性炎症，脓肿大小不一，细菌定植，肝脏和肾脏渗出物中含有中性粒细胞和巨噬细胞。

诊断方法

（1）分离培养

①迟缓爱德华氏菌：使用胰胨大豆琼脂平板，25℃培养24～48h，可形成直径0.5～1mm的圆形、灰白色、湿润、有光泽、隆起的半透明菌落。

②鮰爱德华氏菌：使用胰胨大豆琼脂平板，28 ~ 36℃培养48h，可形成针尖大小的菌落。

（2）生化鉴定 可根据《伯杰氏手册》或API细菌鉴定系统、全自动微生物鉴定系统进行生化鉴定，主要生化特性为革兰氏阴性，不分解蔗糖、甘露醇、乳糖、尿素，葡萄糖产酸产气试验阳性，V-P反应阴性，MRS试验阳性，靛基质阳性，色氨酸脱羧酶试验阳性，精氨酸脱羧酶阴性，西蒙氏柠檬酸盐阴性，不产H_2S。

（3）16S rDNA鉴定 引物F（5′-AGA-GTT-TGA-TCC-TGG-CTC-AG-3′）和引物R（5′-ACG-GCT-ACC-TTG-TTA-CGA-CTT-3′），退火温度为55℃，扩增产物长度为1 407bp。

扩增产物测序后判定。

防治方法

（1）保持水质优良及稳定，加强饲料管理，泼洒益生菌，避免养殖密度过高，投喂营养全面、优质的饲料，增强鱼体抵抗力，保持适度的水温。

（2）漂白粉浓度为1 ~ 1.2mg/L，全池泼洒。

（3）在水温23℃时，饵料中添加适量的维生素C，可增加斑点叉尾鮰对迟缓爱德华氏菌的抵抗力。

（4）用福尔马林灭活的迟缓爱德华氏菌和其胞外及胞内产物作为疫苗，采用口服、注射、浸泡3种方法对牙鲆进行免疫。

Edwardsiellosis

Disease overview

[Disease Characteristic] Severe infectious disease of fish. *Edwardsiella tarta* mainly causes emphysematous putrefactive disease in catfish (EPDC). *Edwardsiella ictaluri* mainly causes enteric septicemia in catfish (ESC). Both can also cause red-head disease of yellow catfish.

[Susceptible Host]

(1) *Edwardsiella tarta* mainly affect various freshwater and marine fish species including Japanese eel (*Anguilla japonica*), olive flounder (*Paralichthys olivaceus*), *Lates calcarifer*, channel catfish (*Ictalurus punctatus*), largemouth bass (*Micropterus salmoides*), flathead grey mullet (*Mugil cephalus*), crimson seabream (*Evynnis japonica*), Nile tilapia (*Tilapia nilotica*), Chinook salmon (*Oncorhynchus tshawytscha*), red seabream (*Pagrus major*), yellowtail (*Seriola lalandi*), rainbow trout (*Oncorhynchus mykiss*), common carp (*Cyprinus carpio*), European bass (*Dicentrarchus labrax*), turbot (*Scophthalmus maximus*), walking catfish (*Clarias batrachus*), brook trout (*Salvelinus fontinalis*), catla (*Catla catla*), roho labeo (*Labeo rohita*), European eel

(*Anguilla anguilla*), Amur catfish (*Silurus asotus*), etc. In addition, this bacterium can also infect invertebrates, amphibians, reptiles, birds and mammals.

(2) *Edwardsiella ictaluri* mainly affect catfish including channel catfish, white bullhead, walking cafish and brown bullhead, of which channel catfish is the most susceptible. It can also infect other fish species such as Japanese eel and ornamental fish such as zebrafish, rosy barb, bronze featherback. Artificial inoculation can infect Chinook salmon and rainbow trout.

[Susceptible Stage] *Edwardsiella tarta* mainly affects fish over 450g, while *Edwardsiella ictaluri* can affect fish of any age.

[Outbreak Water Temperature] Disease outbreaks can occur throughout the year at a water temperature of 15℃ . The peak of outbreaks usually occurs at a water temperature of 25~30℃ . EPDC outbreak seldom occurs over 30℃ . ESC outbreaks occur at a water temperature of 20~30℃ or even lower. Red-head disease of yellow catfish outbreaks occur at a water temperature at 18~30℃ with endemic outbreaks at 20~28℃ , can even occur over 30℃ with worsened environmental conditions.

[Geographical Distribution] Widely distributed globally, mainly in Australia, India, the Malay Archipelago, Israel, Japan, Panama, USA, as well as other tropical and subtropical regions. The disease causes particularly great impact to the eel aquaculture industry in Japan and Chinese Taipei.

Aetiological agents

(1) *Edwardsiella tarda* and *Edwardsiella ictaluri*, of which *Edwardsiella tarda* is a zoonotic pathogen.

(2) Family: Enterobacteriaceae. Genus: *Edwardsiella*.

(2)Opportunistic pathogen. Gram-negative short rod bacterium, facultative anaerobic, (0.5~1) μm × (1~3)μm in size, non-spore forming, non-capsulated, covered with flagella and motile, but those isolated from red seabream and herring are non-motile.

(4) Able to grow from 15~42 ℃ . The optimal growing temperature for *Edwardsiella tarda* and *Edwardsiella ictaluri* are 37℃ and 25~30℃ , respectively.

(5) *Edwardsiella tarda* possesses O antigen on the bacterial body and H antigen on the flagella. According to the respective O antigens, it can be sub-categorized into 4 serotypes: A, B, C, and D, of which serotype A has the highest pathogenicity. For *Edwardsiella ictaluri*, it is generally believed that only one serotype is present, but in other reports there are of at least 2 different serotypes or genotypes according to the sources of the strains (catfish or non-catfish).

Clinical signs and pathological changes

(1) Fish affected by *Edwardsiella tarda* only show 3~5mm skin wounds on the body surface at the initial stage of infection. In severe case, ulcerations are present in the body flank and the base of the tail, which expand rapidly forming a hollow structure which appears as emphysema. Discoloration is usually seen in these lesions, which release smelly odor upon dissection. Necrotic

tissues occupy around 1/3 of these hollow structures.

(2) *Edwardsiella ictaluri* can cause chronic infection and acute infection:

① Chronic infection: Acquired via infection of the nervous tissues. It causes granulomatous inflammation leading to chronic meningitis. Affected fish are inactive with decreased appetite; float on the water surface, sometimes in a hanging posture of "head-up-tail-down" and accompanied with spiraling and abnormal rapid swimming motions. The adjacent areas of the lesions show discoloration with interocular vertical ulceration, which further progresses and exposes the underlying bones resulting in an open ulcer on the forehead.

② Acute infection: Acquired via gastrointestinal tract and enter bloodstream. Affected fish show abdominal distention with tiny hyperaemic, hemorrhagic and exudative lesions on the body surface or muscles. The ascites are not easy to clot. Hepatic oedema, being fragile in texture with petechiae and tiny greyish white necrotic foci. Spleen enlargement and haemorrhage. Gastrointestinal distension, hyperaemia and inflammation, with mucosal oedema, hyperaemia and lumens filled gas and yellowish fluid.

(3) Yellow catfish affected by red-head disease are segregated, show decreased appetite, abdominal distension with hyperaemia and haemorrhages present in fin base, submandibular, operculum and the abdomen as well as anal protrusion. In post-mortem examination, abundant bloody or clear fluid is present inside the abdomen with liver enlargement (petechiae or ecchymosis), kidney enlargement (hyperaemic), splenic enlargement (appearing dark-red), and intestinal distension (hyperaemic and inflamed, with abundant yellowish fluid inside the lumens).

(4) Histologically, interstitial nephritis, purulent hepatitis and purulent splenitis are the major characteristics. The abscesses are non-uniform in size with bacterial colonization. The exudates of liver and kidney contain neutrophils and macrophages.

Diagnostic methods

(1) **Bacterial isolation**

① *Edwardsiella tarda*: Inoculate sample on tryptic soy agar (TSA) plate and incubate at 25℃ for 24~48 hours. It forms round, greyish white, mucoid, shinny and convex translucent colonies of approximately 0.5~1mm in diameter.

② *Edwardsiella ictaluri*: Inoculate sample on tryptic soy agar (TSA) plate and incubate at 28~36℃ for 48 hours. It forms pinpoint size colonies.

(2) **Biochemical identification** With reference to *Bergey's Manual* or using API identification system or automatic microbial indentification system to conduct microbiology biochemical identification. Major biochemical characteristics are: gram-negative; non-fermenter for sucrose, mannitol, galactose and urease; glucose fermenter; V-P reaction negative; produce gas in MRS liquid medium; indole test positive; TDC test positive; ADC test negative; Simmons citrate test negative; non-H_2S producing.

(3) **16S rDNA sequencing identification** Use the forward primer (5′-AGA-GTT-TGA-

TCC-TGG-CTC-AG-3′) and reverse primer (5′-ACG-GCT-ACC-TTG-TTA-CGA-CTT-3′) with annealing temperature at 55℃ to perform PCR. The amplicon size is 1,407bp. Perform sequencing on the amplicon for confirmation.

Preventive measures

(1) Maintain good and stable water quality; strengthen feed management; apply probiotics; avoid excessive cultivation density; provide all-round nutritional and high-quality feed; boost the immunity of the fish; and provide a suitable water temperature.

(2) Bleaching powder: 1~1.2mg/L, apply to the whole pond.

(3) Apply an optimal amount of vitamin C into the feed when the water temperature is 23℃ , which can increase channel catfish's tolerance against *Edwardsiella tarda*.

(4) Prepare inactivated vaccine using formalin-inactivated *Edwardsiella tarda* and its extracellular and intracellular products. Vaccinate olive flounder by oral, injection and immersion routes for immunization.

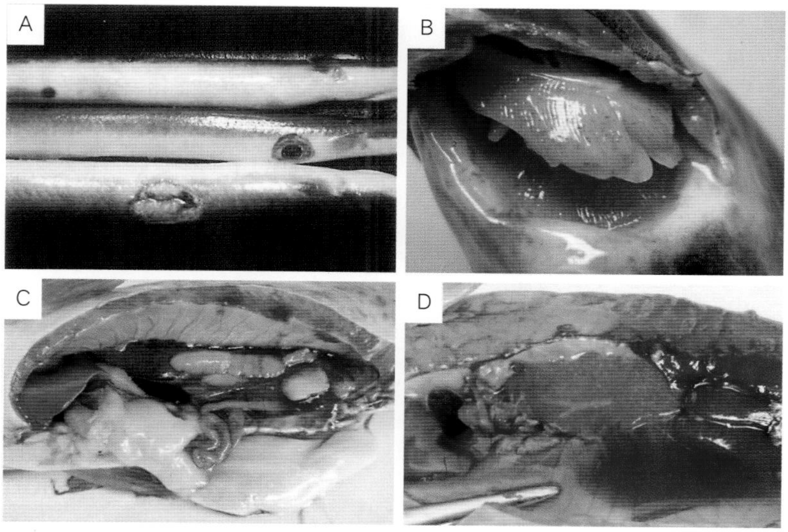

感染爱德华氏菌的病鱼

A．体表皮肤呈现溃烂、穿孔（位于肝脏位置）

B．鳃丝上有大量出血点

C．肝脏和肠道出血，腹水

D．肾脏出血，腹水

[源自唐绍林及雷燕]

Macroscopic findings of affected fish

A．Ulcerations and perforations present on the body truck

B．Petechiae on the gill

C．Pale liver, intestinal haemorrhage and ascites

D．Renal haemorrhage and ascites

[Source：Shaolin Tang and Yan Lei]

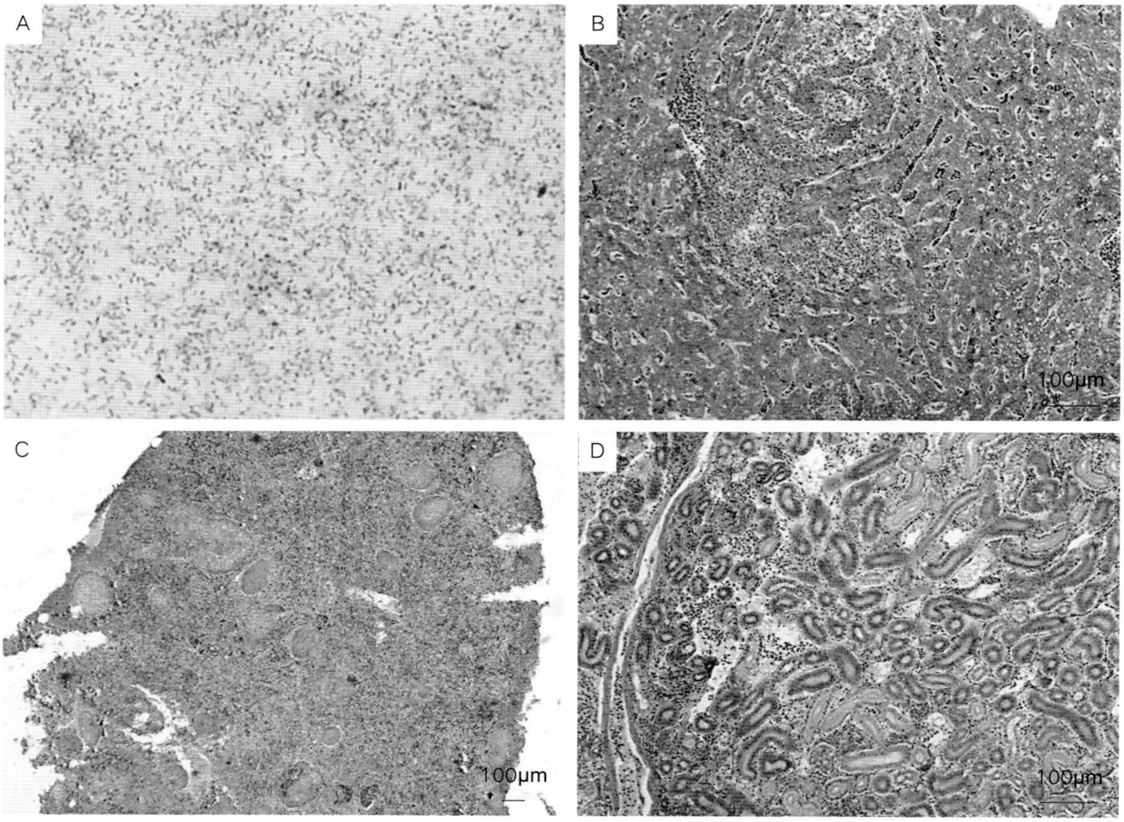

病鱼细胞及组织病变

A．革兰氏染色后的迟缓爱德华氏菌（革兰氏染色）

B．肝脏瘀血，肝细胞固缩，有坏死区（HE染色）

C．脾脏多坏死灶，可见黑色素巨噬细胞中心（HE染色）

D．肾脏瘀血、出血，肾小球瘀血，有坏死，肾间组织细胞零星坏死（HE染色）

[源自彭天辉及唐绍林]

Histological lesions of affected fish

A．Gram-negative short rod-shaped *Edwardsiella tarda* (Gram staining)

B．Liver. Focal area of hepatocellular pyknosis and necrosis (HE staining)

C．Spleen. Multifocal areas of necrosis (HE staining)

D．Kidney. Renal haemorrhage and sporadic interstitial necrosis (HE staining)

[Source：Tianhui Peng and Shaolin Tang]

鱼立克次氏体病

疾病概述

【概述】 鱼立克次氏体病又称鲑鱼立克次氏体败血症、银大麻哈鱼综合征，是一种可导致高死亡率和巨大经济损失的流行病。

【宿主】 主要感染银大麻哈鱼（*Oncorhynchus kisutch*）、大鳞大麻哈鱼（*O. tshawytscha*）、虹鳟（*O. mykiss*）、细鳞大麻哈鱼（*O. gorbuscha*）、马苏大麻哈鱼（*O. masou*）、大西洋鲑（*Salmo salar*）、花鲈（*Lateolabrax japonicus*）、莫桑比克罗非鱼（*Oreochromis mossambicus*）等。

【易感阶段】 各阶段易感。

【发病水温】 15 ～ 18℃为发病高峰期，死亡率差异很大，从智利（银大麻哈鱼）90%的高发率，至加拿大和挪威0.06%的低发率。

【地域分布】 1989年，首次在智利饲养殖的银大麻哈鱼中发现，主要流行于智利、加拿大（东、西海岸）、挪威、爱尔兰。

病原

（1）病原为鲑鱼立克次氏体（*Piscirickettsia salmonis*）。

（2）属鱼立克次氏体科（Piscirickettsiaceae）、鱼立克次氏体属（*Piscirickettsia*）。

（3）多形性，主要是球形，直径0.5 ～ 1.5μm，专性细胞内增生，在宿主组织细胞有膜结合的胞质空泡内复制，革兰氏阴性菌。

（4）对红霉素、噁喹酸、四环素等一系列抗生素敏感，但对青霉素不敏感。

（5）可在敏感细胞如CHSE-214、EPC、CHH等培养增殖，产生CPE，滴度达10^6 ～ 10^7 $TCID_{50}$/mL，但不能在任何已知的无细胞培养基中增殖。最适培养温度为15 ～ 18℃，在20℃以上或10℃以下延缓生长，在25℃以上不生长。

（6）鲑鱼立克次氏体的致病物质主要有内毒素和磷脂酶A两类。其中，内毒素可导致内皮细胞损伤和微循环障碍等病理变化；磷脂酶A能溶解宿主细胞膜或吞噬体膜。

临床症状和病理学变化

（1）常聚集在网箱的边缘水面上，昏睡，体色发黑，食欲减退。

（2）患病早期，体表出现小白病灶或出血性溃疡。

（3）鳃灰白，血细胞压积为25%或更少。

（4）全身多个器官和组织受侵害，肾脏肿胀、变色，脾脏肿大，腹腔积水，内脏脂肪、胃、鱼鳔和肌肉出血，更严重的感染鱼肝脏苍白，并可能会出现奶油色、直径5 ～ 6mm的圆形不透明结节。

(5) 组织学上可见鳃呈多灶性上皮增生，伴轻度至中度增生组织坏死，在严重固化的二级鳃片区域有细菌存在。真皮和表皮坏死，皮下肌肉组织变性。肠道组织受损严重，上皮细胞坏死脱落。肾脏有炎症细胞浸润，造血组织广泛坏死，伴水肿和纤维化。

诊断方法

(1) 组织病理学检查 脏器中的巨噬细胞胞质内发现许多嗜碱性或多染性（HE染色），或是暗蓝色直径约1μm的椭圆形小体。

(2) 细菌分离 使用CHSE-214或EPC细胞系，培养温度为15 ~ 18℃。

(3) PCR PS2S (223F) (5′-TAG-GAG-ATG-ATG-AGC-CCG-CGT-TG-3′) 和PS2AS (690R) (5′-GCT-ACA-CCT-GCG-AAA-CCA-CTT-3′)，退火温度为50℃，扩增产物长度为476bp。

扩增产物经测序后判定。

(4) 免疫荧光或免疫组化

防治方法

(1) 减少应激，如减少鱼密度、渔网变换次数和分级次数等。

(2) 及时从网箱中清除死鱼和垂死鱼，减少每个站点鱼数目，留下一些水面休渔一段时间。

(3) 每千克鱼体重口服30 ~ 50mg土霉素，可降低死亡率。

(4) 鲑鱼立克次氏体细胞培养上清裂解液1 : 2稀释，经福尔马林灭活制成菌苗，可对银大麻哈鱼起保护作用。

Piscirickettsiosis

Disease overview

[Disease Characteristic] Epidemic disease causing high mortality rate and huge economic loss in the aquaculture industry. Also known as ricketts septicemia of salmon and silver salmon syndrome.

[Susceptible Host] Mainly affects silver salmon (*Oncorhynchus kisutch*), Chinook salmon (*O. tshawytscha*), rainbow trout (*O. mykiss*), pink salmon (*O. gorbuscha*), masu salmon (*O. masou*), Atlantic salmon (*Salmo salar*), sea bass (*Lateolabrax japonicas*), Mozambique tilapia (*Oreochromis mossambicus*), etc.

[Susceptible Stage] Fish are susceptible at all stages.

[Outbreak Water Temperature] Peak of disease outbreak occurs at 15~18℃ , with varying mortality rates in silver salmon from as high as 90% in Chile to as low as 0.06% in Canada and Norway.

[Geographical Distribution] First reported in a silver salmon aquaculture establishment in Chile in 1989. Disease mainly endemic in Chile, Canada (East and West Coast), Norway and Ireland.

Aetiological agent

(1) *Piscirickettsia salmonis*.

(2) Family: Piscirickettsiaceae. Genus: *Piscirickettsia*.

(3) Polymorphic, mainly spherical, with a diameter of 0.5~1.5μm, obligate intracellular organism which replicates in the cytoplasmic vacuoles of membrane-bound cells of the host. Gram-negative bacteria.

(4) Sensitive to a range of antibiotics such as erythromycin, oxaquine, and tetracycline, but not sensitive to penicillin.

(5) It can be cultured in susceptible cell lines such as CHSE-214, EPC, CHH, etc. to produce CPE with a titer reaching 10^6~10^7 $TCID_{50}$/mL, but it cannot grow in any known cell-free culture media. The optimum incubation temperature is 15~18℃ , with slower growth above 20℃ or below 10℃ , and stop growing above 25℃ .

(6) The pathogenic factors of *Piscirickettsia salmonis* are mainly endotoxin and phospholipase A. Endotoxin can cause pathological changes such as endothelial cell damage and microcirculation disorders; while phospholipase A can lyse the membranes of host cells or phagosome.

Clinical signs and pathological changes

(1) Affected fish often clustering at the water surface near the edge of the net cages, with signs of lethargy, darkened body color, and reduced appetite.

(2) In the early stage of disease, the body surface has small white lesions or hemorrhagic ulceration.

(3) The gills appear grayish white, with packed cell volume at 25% or less.

(4) Multiple organs and tissues of the whole body are affected: kidney swelling and discoloration, splenomegaly, ascites, haemorrhages of visceral fat, stomach, swim bladder and muscles. In severe cases, infected fish shows pale liver, which may possess creamy white, round, opaque nodules, with diameters of 5~6mm.

(5) Histologically, multifocal epithelial hyperplasia is present in the gills, accompanied with mild to moderate necrosis of the hyperplastic tissues. In severely consolidated secondary lamella,

bacteria may be present. Necrosis is also present in the dermis and epidermis, with degeneration of subcutaneous muscular tissues. The intestinal tissue is severely damaged, with necrosis and sloughing of epithelial cells. The kidneys are infiltrated with inflammatory cells, the hemopoietic tissues are diffusely necrotic, accompanied with oedema and fibrosis.

Diagnostic methods

(1) **Histopathological examination** Cytoplasm of the macrophages of affected fish contains a lot of basophilic or polychromic (HE staining), or dark blue elliptical bodies with a diameter of around 1μm.

(2) **Bacterial isolation** Use CHSE-214 or EPC cell line and incubate at 15~18℃.

(3) **Polymerase chain reaction (PCR)** The primers are PS2S(223F) (5′-TAG-GAG-ATG-ATG-AGC-CCG-CGT-TG-3′) and PS2AS(690R) (5′-GCT-ACA-CCT-GCG-AAA-CCA-CTT-3′) with annealing temperature at 50℃. The amplicon size is 476bp. Perform sequencing on the amplicon for confirmation.

(4) **Immunofluorescence or immunohistochemistry**

Preventive measures

(1) Reduce stress such as decreasing the cultivation density, the frequency of changing cage nets and the frequency of sorting procedures.

(2) Remove dead and dying fish from net cages as soon as possible. Reduce the quantity of fish at each station, leaving some waterbody free to suspend farming for a period of time.

(3) Oxytetracycline: Apply orally 30~50mg/kg of fish can reduce the mortality rate.

(4) Prepare vaccine by diluting *Piscirickettsia salmonis* cell culture supernatant in 1 : 2 and inactivating by formalin. Subsequent vaccination can provide protection for silver salmon.

感染类立克次氏体的罗非鱼

A．鳃丝上黄白色结节

B．正常鳃丝呈现红白相间

C．鳃及脾脏密发大小不一的黄白色结节

D．肾脏密发大小不一黄白色结节

[源自《水产动物疾病防治及正确用药手册》，李建霖]

Macroscopic findings of affected tilapia

A．Gill. Large number of white nodules is present in the gill filaments

B．Normal gill

C．Numerous white nodules present in the gill and spleen

D．Kidney. Numerous white nodules are present

[Source：*Handbook of Aquatic Animal Disease Control and Medication*，Jianlin Li]

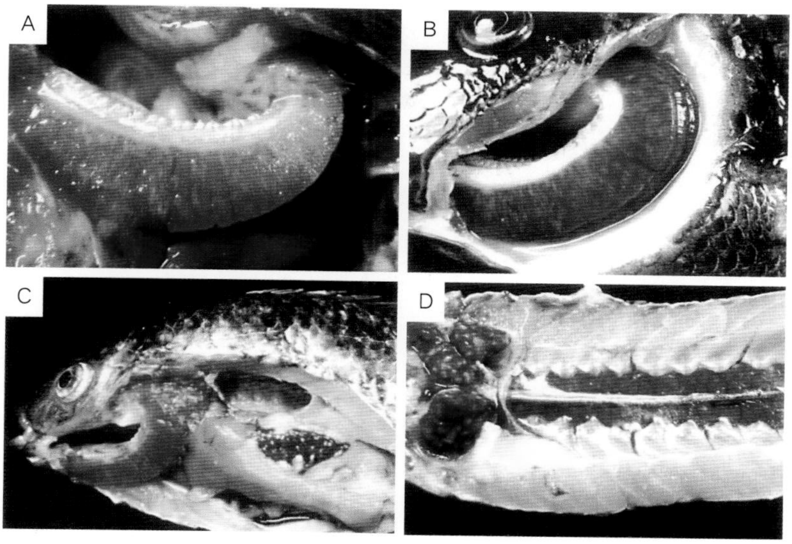

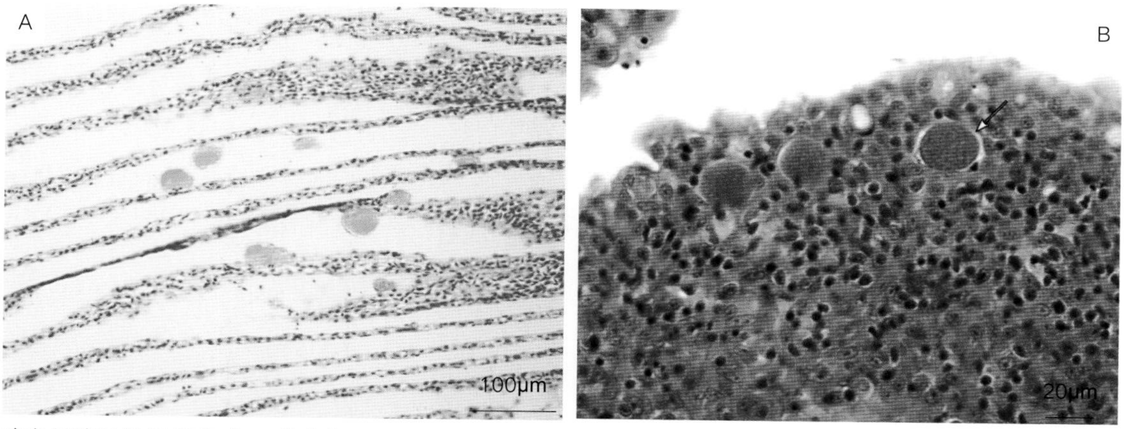

病鱼细胞及组织病变（HE染色）

鳃上皮细胞中可见类立克次氏体　B．组织中的类立克次氏体（箭头）

[源自S．W．Feist及D．Bucke]

Histological lesions of affected fish (HE staining)

A．Rickettsia-like organism within distended gill epithelial cells　B．Rickettsia-like organism within tissue (arrow)

[Source：S．W．Feist and D．Bucke]

淡水鱼细菌性败血病

疾病概述

【概述】 淡水鱼细菌性败血病是一种急性传染病。

【宿主】 主要感染白鲫（*Carassius auratus cuvieri*）、鲫（*Carassius auratus*）、异育银鲫（*Carassius auratus gibelio*）、团头鲂（*Megalobrama amblycephala*）、鲢（*Hypophthalmichthys molitrix*）、鳙（*Hypophthalmichthys nobilis*）、鲤（*Cyprinus carpio*）、鲮（*Cirrhinus molitorella*）及草鱼（*Ctenopharyngodon idella*）、青鱼（*Mylopharyngodon piceus*）等。

【易感阶段】 从夏花鱼种到成鱼均可感染，以2龄成鱼为主。

【发病水温】 水温9 ~ 36℃均有流行，尤以水温持续在28℃以上及高温季节后水温仍保持在25℃以上时较为严重，死亡率高达60% ~ 100%。

【分布地域】 全世界广泛分布。

病原

（1）病原菌有嗜水气单胞菌（*Aeromonas hydrophila*）、温和气单胞菌（*Aeromonas sobria*）等多种细菌。

（2）均属气单胞菌科（Aeromonadaceae）、气单胞菌属（*Aeromonas*）。

（3）嗜水气单胞菌属细菌为革兰氏阴性菌，呈杆状，两端钝圆，中轴端直，大小为（0.5 ~ 0.9）μm×（1.0 ~ 2.0）μm，单个散在或两两相连，极端单鞭毛，有运动性，无芽孢、无荚膜；最适生长pH为7.27，最适盐度为0.05%，最适生长温度为22 ~ 28℃，不产生色素。

（4）温和气单胞菌属细菌为革兰氏阴性菌，兼性厌氧菌，呈短杆状，两端呈钝圆，直径0.3 ~ 1.0μm，长度1.0 ~ 3.5μm。具有长在菌体一端的鞭毛，有运动性；最适生长温度为28℃。

临床症状和病理学变化

（1）厌食或不吃食，静止不动或阵发性乱游、乱窜，有的在池边摩擦，最后衰竭死亡。

（2）急性感染早期，体表和口腔等部位轻度充血。严重时，鱼体表严重充血，甚至出血，眼球突出，眼眶周围也出血，肛门红肿，腹部膨大，腹腔内积有淡黄色透明腹水或红色混浊腹水。

（3）红细胞肿大，有溶血现象发生，胞浆内出现大量嗜伊红颗粒、胞浆透明化，在脾、肝、胰、肾中均有较多血源性色素沉着。

诊断方法

（1）细菌分离 使用普通营养琼脂，37℃培养24 ～ 48h，菌落圆形，灰白色，半透明，表面光滑湿润，微凸，边缘整齐。

（2）生化鉴定 可根据《伯杰氏手册》或API细菌鉴定系统、全自动微生物系统进行生化鉴定，主要生化特性为：吲哚、氧化酶试验阳性，葡萄糖、麦芽糖、蔗糖、甘露糖、阿拉伯糖、七叶苷、水杨酸等糖发酵试验呈阳性，不能利用乳糖和鸟苷酸。

（3）16S rDNA鉴定 引物为：F（5′-GGG-AGT-GCC-TTC-GGG-AAT-CAG-A-3′）和R（5′-TCA-CCG-CAA-CAT-TCT-GAT-TTG-3′），退火温度为63℃，扩增产物长度为356bp。

扩增产物经测序后判定。

防治方法

（1）冬季干塘彻底清淤，并用生石灰或漂白粉彻底消毒，以改善水体生态环境。

（2）发病鱼池用过的工具要进行消毒，病死鱼要及时捞出深埋。

（3）鱼种尽量就地培养，减少搬运，并注意下塘前进行鱼体消毒。可用15 ～ 20mg/L的高锰酸钾溶液药浴10 ～ 30min。

（4）流行季节，用25 ～ 30mg/L的生石灰化浆全池泼洒，每半个月1次，以调节水质。食场定期用漂白粉、漂白粉精等进行消毒。

（5）漂白粉1mg/L，漂白粉精（有效氯60％～ 65％）0.2 ～ 0.3mg/L，二氧化氯0.1 ～ 0.3mg/L或二氯海因0.2 ～ 0.3mg/L，全池泼洒以预防。

（6）每千克鱼体重用氟苯尼考5 ～ 15mg制成药饵投喂，每天1次，连用3 ～ 5d。

（7）复方新诺明第一天100mg，第二天开始药量减半，拌在饲料中投喂，5d为一个疗程。

Bacterial septicemia of freshwater fish

Disease overview

[Disease Characteristic] Bacterial Septicemia of freshwater fish is an acute infectious disease.

[Susceptible Host] Mainly affect white crucian carp (*Carassius auratus cuvieri*), goldfish (*Carassius auratus*), Prussian carp (*Carassius auratus gibelio*), Wuchang bream (*Megalobrama amblycephala*), silver carp (*Hypophthalmichthys molitrix*), bighead carp (*Hypophthalmichthys nobilis*), common carp (*Cyprinus carpio*), mud carp (*Cirrhinus molitorella*), grass carp

(*Ctenopharyngodon idella*) and black carp (*Mylopharyngodon piceus*).

[Susceptible Stage] Fish are susceptible from juvenile stage in summer to adult stage, mainly affecting adult fish of 2-year-old.

[Outbreak Water Temperature] Disease outbreaks can occur at water temperature of 9~36℃, with particularly severe outbreaks occurring at a water temperature persistently maintaining at 28℃ or above and after warm season with water temperature above 25℃. Mortality rate can reach 60%~100%.

[Geographic Distribution] Widely distributed globally.

Aetiological agents

(1) Mainly *Aeromonas* spp. including *Aeromonas hydrophila* and *Aeromonas sobria*.

(2) Family: Aeromonadaceae. Genus: *Aeromonas*.

(3) *Aeromonas hydrophila*: Gram-negative short-rod bacteria, obtuse at both ends with straight trunk, approximately (0.5~0.9)μm × (1.0~2.0)μm in size, arranged singly or in pairs, with single polar flagella, motile, non-spore forming and non-capsulated. Optimal growing pH at 7.27; optimal salinity at 0.05%; optimal growing temperature at 22~28℃ without any pigmentation.

(4) *Aeromonas sobria*: Gram-negative short-rod bacteria, facultative anaerobic, obtuse at both ends, 0.3~1.0μm in diameter, 1.0~3.5μm in length, with single polar flagella, motile, optimal growing temperature at 28℃ .

Clinical signs and pathological changes

(1) Anorexia, lethargic or intermittent abnormal rapid swimming movement, sometimes rubbing against the pond side, eventually died of weakness.

(2) At early stage of acute infection, mild hyperaemia is present on the body surface and the oral cavity. In severe case, the body surface of affect fish is severely hyperaemic or haemorrhagic, with exophthalmia, periocular haemorrhage, anal reddening and swelling, abdominal distension, yellowish clear ascites or bloody turbid ascites inside the abdomen.

(3) Marked swelling of red blood cells resulting in hemolysis, with abundant eosinophilic granules inside the cytoplasm, which appears translucent. Haematogenous pigmentations are present in the spleen, liver, pancreas and kidney.

Diagnostic methods

(1) **Bacterial isolation** Use normal nutritional agar to incubate at 37 ℃ for 24~48 hours. The bacterial colonies are round, greyish white in color, translucent, mucoid, slightly convex and with clear margins.

(2) **Biochemical identification** With reference to *Bergey's Manual* or using API

identification system or automatic microbial identification system to conduct microbiology biochemical identification. Major biochemical characteristics are: indole and oxidase tests positive, fermenting glucose, maltose, sucrose, mannose, arabinose, esculin and citric acid, but not galactose and GMP.

(3) **16S rDNA identification** Perform PCR amplification using the forward primer (5′-GGG-AGT-GCC-TTC-GGG-AAT-CAG-A-3′) and reverse primer (5′-TCA-CCG-CAA-CAT-TCT-GAT-TTG-3′) with annealing temperature at 63℃. The amplicon size is 356bp. Perform sequencing on the amplicon for confirmation.

Preventative measures

(1) In winter, completely stamp out the culture pond to remove sludge and thoroughly disinfect with quicklime or bleaching powder to improve the ecological environment of the waterbodies.

(2) Disinfect the equipment used in infected pond. Remove dead fish as soon as possible for burial.

(3) Avoid unnecessary transportation of fingerlings and implement disinfection prior to stocking of new fish, which can be done by immersion in 15~20mg/L sodium permanganate solution for 10~30 min.

(4) During outbreak season, apply 25~30mg/L quicklime to the culture pond once every half-month to maintain the water quality. Disinfect the feeding site with bleach regularly.

(5) Apply 1mg/L bleaching powder, bleaching powder concentrate (60%~65% available chlorine), 0.1~0.3mg/L chlorine dioxide or 0.2~0.3mg/L Dichlorohydantoin (DCDMH) to the whole culture pond for prevention.

(6) For every one kilogram of fish, apply 5~15mg florfenicol in feed once a day and treat for consecutive 3~5 days.

(7) For every one kilogram of fish, apply 100mg trimethoprim/sulfamethoxazole in feed on the first treatment day and then with halved dose on the second day onwards for a complete treatment course (5 consecutive days).

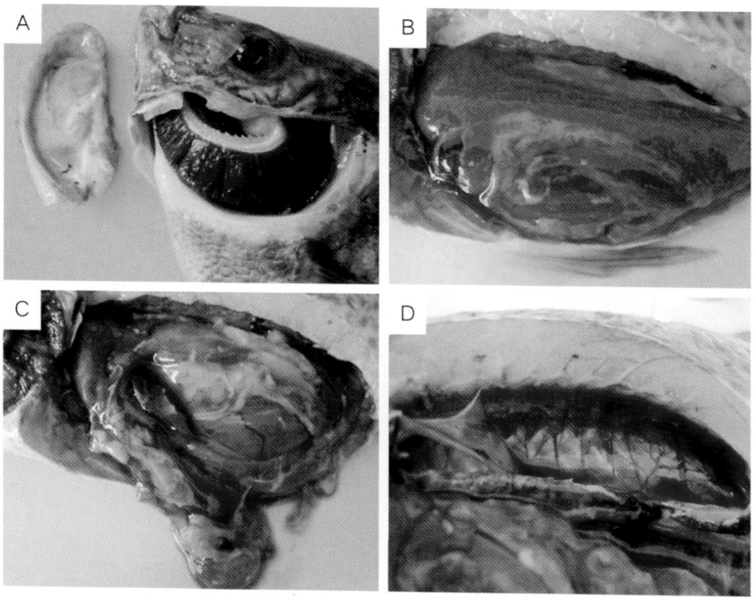

嗜水气单胞菌引起的罗非鱼细菌性败血症

A．病鱼烂鳃，鳃上附着泥
B．腹水，花肝，空肠，出血，胆囊肿大
C．脾脏肿大，呈暗黑色 D．肾脏出血
[源自雷燕]

Macroscopic findings of Tilapia fish infected by *Aeromonas hydrophila*

A．Marked hyperaemia and necrosis of gill B．Patchy necrosis of the liver with gall bladder enlargement. Empty intestines, with ascites and haemorrhage C．Enlarged spleen, appearing dark red in color D．Marked haemorrhage of kidney
[Source：Yan Lei]

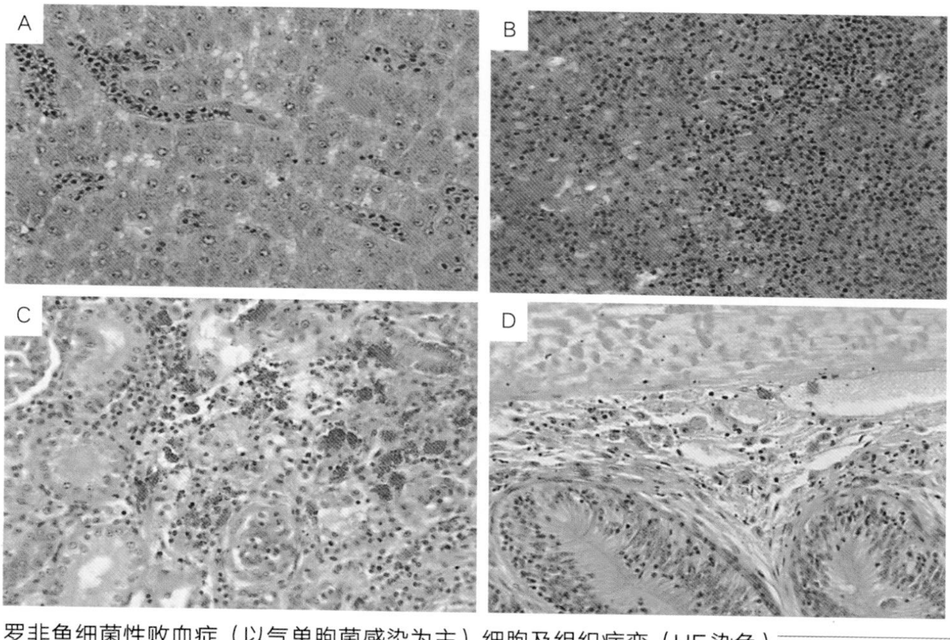

罗非鱼细菌性败血症（以气单胞菌感染为主）细胞及组织病变（HE染色）

A．肝瘀血严重，轻微水变，肝细胞界限不明显 B．脾瘀血严重，细胞坏死 C．肾组织中有大量的嗜酸性粒细胞，肾间组织细胞有坏死 D．肠瘀血，细胞坏死，肠腔空
[源自戚瑞荣]

Histological lesions of Tilapia fish infected by *Aeromonas* sp. (HE staining)

A．Liver. Marked congestion and mild hydropic degeneration of the hepatocytes. The margins of the hepatocytes become indistinct B．Spleen. Congestion and mild necrosis C．Kidney. Numerous eosinophils are present with mild necrosis of the interstitial tissues D．Intestine. Mild necrosis is present in the lamina propria of the mucosa
[Source：Ruirong Qi]

细菌性肠炎病

疾病概述

【概述】 细菌性肠炎病是一种肠道致病菌所引起的传染性疾病，是养殖鱼中严重的疾病之一。

【宿主】 主要感染草鱼（*Ctenopharyngodon idellus*）、青鱼（*Mylopharyngodon piceus*）、鲤（*Cyprinus carpio*）、鳙（*Aristichthys nobilis*）。

【易感阶段】 青鱼、草鱼从鱼种至成鱼都可受害，死亡率高，一般死亡率在50%左右，发病严重的鱼池死亡率可高达90%。

【发病水温】 流行高峰水温为25 ～ 30℃。

【地域分布】 全世界广泛分布。

病原

（1）病原为肠型点状气单胞菌（*Aeromonas punotata* f. *intestinalis*）。

（2）属气单胞菌科（Aeromonadaceae）、气单胞菌属（*Aeromonas*）。

（3）革兰氏阴性短杆菌，两端钝圆，多数两个相连。极端单鞭毛，有运动能力，无芽孢。大小为（0.4 ～ 0.5）μm×（1 ～ 1.3）μm。

（4）在RS选择和鉴别培养基上，菌落呈黄色。pH 6 ～ 12均可生长。生长适宜温度为25℃，在60℃下30min则死亡。

临床症状和病理学变化

（1）病鱼离群独游，游动缓慢，体色发黑，食欲减退以至完全不吃食。

（2）病情较重的，腹部膨大，两侧有红斑，肛门常红肿外突，呈紫红色，轻压腹部，有黄色黏液或血脓从肛门处流出。有的病鱼仅将头部提起，即有黄色黏液从肛门流出。

（3）早期可见肠壁充血发红、肿胀发炎，肠腔内没有食物或只在肠的后段有少量食物，肠内有较多黄色或黄红色黏液。疾病后期，可见全肠充血发炎，肠壁呈红色或紫红色，尤其以后肠段明显，肠黏膜往往溃烂脱落，并与血液混合而成血脓，充塞于肠管中。病情严重的，腹腔内常有淡黄色腹水，腹壁上有红斑，肝脏常有红色斑点状瘀血。

（4）病理组织学观察可见固有层内毛细血管显著充血、出血，肠黏膜上皮变性、脱落，严重者黏膜上皮解体，裸露出严重充血、出血的固有层，肠腔内有大量炎性分泌物。

诊断方法

（1）分离培养 使用胰胨大豆琼脂（TSA）平板，25℃培养24 ～ 48h，菌落周围可产生褐色色素，半透明。

（2）生化鉴定 可根据《伯杰氏手册》或API细菌鉴定系统、全自动微生物系统进行生化鉴定，主要生化特性为：分解蔗糖、甘露醇、水杨酸、七叶苷，不分解枸橼酸盐，细胞色素氧化酶试验阳性，发酵葡萄糖产酸产气或产酸不产气，吲哚试验阳性，硝酸盐还原阳性。

（3）16S rDNA鉴定

正向引物：63F：5′-CAG-GCC-TAA-CAC-ATG-CAA-GTC-3′。

反向引物：1 387R：5′-GGG-CGG-WGT-GTA-CAA-GGC-3′，退火温度为52℃，扩增产物长度为1 280bp。

扩增产物经测序后判定。

防治方法

（1）彻底清塘消毒，保持水质清洁。严格执行“四消（菌种消毒、饵料消毒、工具消毒、食场消毒）、四定（定质、定量、定时、定位）”措施。投喂新鲜饲料，不喂变质饲料。

（2）鱼种放养前，用8 ～ 10mg/L浓度的漂白粉浸洗15 ～ 30min。

（3）发病季节，每隔15d用漂白粉或生石灰在食场周围泼洒消毒；或用浓度为1mg/L的漂白粉或20 ～ 30mg/L的生石灰全池泼洒，消毒池水，可控制此病发生。发病时可用以上任一药物每天泼洒，连用3d。

（4）每千克鱼体重每天用大蒜（用时捣烂）5g或大蒜素0.02g、食盐0.5g，拌饲料分上、下午2次投喂，连喂3d。

（5）每千克鱼体重每天用干的地锦草、马齿苋、铁苋菜或辣蓼（合用或单用均可）各5g（打成粉），食盐0.5g，拌饲料分上、下午2次投喂，连喂3d。如用新鲜的，则地锦草、马齿苋为25g，铁苋菜、辣蓼为20g。

（6）每千克鱼体重每天用干的穿心莲20g或新鲜的穿心莲30g，打成浆。再加盐0.5g，拌饲料分上、下午2次投喂，连喂3d。

Bacterial enteritis

Disease overview

[Disease Characteristic] Infectious disease caused by intestinal pathogens, a severe disease for fish aquaculture.

[Susceptible Host] Mainly affect grass carp (*Ctenopharyngodon idellus*), black carp (*Mylopharyngodon piceus*), common carp (*Cyprinus carpio*) and bighead carp (*Aristichthys nobilis*).

[Susceptible Stage] Fingerling to adult stages of grass carp and black carp are all susceptible. High mortality rate, usually around 50% which can reach 90% in severe outbreak.

[Outbreak Water Temperature] Disease outbreaks occur at a water temperature of 25~30℃ .

[Geographical Distribution] Widely distributed globally.

Aetiological agent

(1) *Aeromonas punotata* f. *intestinalis*

(2) Family: Aeromonadaceae. Genus: *Aeromonas*.

(3) Gram-negative short rod bacteria, obtuse at both ends and usually paired, (0.4~0.5)μm × (1~1.3)μm in size. The bacteria possess single polar flagella at one side, being motile and non-spore forming.

(4) It forms yellowish colonies on Rimler-Shotts (RS) medium, able to grow at pH 6~12 with optimal growing temperature at 25℃. Cannot survive at 60℃ for 30 minutes.

Clinical signs and pathological changes

(1) Affected fish are segregated, with slow swimming movement, darkened body color, decreased appetite or even anorexia.

(2) In severe case, affect fish show abdominal distension with ecchymosis on the body flank and anal protrusion, which appears purplish red with yellowish mucus or bloody purulent discharge from the anus upon gentle compression or "head-up-tail-down" positioning.

(3) At early stage of infection, affect fish may show intestinal hyperaemia and enteritis with yellowish or yellowish red mucus content. At later stage of infection, intestinal hyperaemia and enteritis extend to the whole gastrointestinal tract, intestinal wall appearing red or purplish red especially the caudal sections. Mucosal ulceration is common, of which the necrotic debris admix with blood forming bloody purulent discharge filling up the intestinal lumens. In severe case, there are usually yellowish ascites in the abdomen, ecchymosis on abdominal wall and petechial congestion in liver.

(4) Histologically, the capillaries in the lamina propria of the intestinal epithelium are apparently hyperaemic or sometimes with haemorrhage. Degeneration of the intestinal mucosa, sloughing off and disintegrate causing ulceration with inflammatory exudates filling up the lumens.

Diagnostic methods

(1) **Bacterial isolation** Use trypticase soy agar and incubate for 24~48h at 25℃. Brownish pigments may be present adjacent to the translucent colonies.

(2) **Biochemical identification** With reference to *Bergey's Manual* or using API identification system or automatic microbial identification system to conduct microbiology biochemical identification. Major biochemical characteristics are: metabolizing sucrose, mannitol, salicylic acid and esculin, but not citric acid; cytochrome oxidase test positive; fermenting glucose to form acid with/without gas; indole test positive; nitrate reduction test positive.

(3) **16S rDNA identification** Perform PCR amplification using the primers 63F (5′-CAG-GCC-TAA-CAC-ATG-CAA-GTC-3′) and 1, 387R (5′-GGG-CGG-WGT-GTA-CAA-GGC-3′) with annealing temperature at 52℃. The amplicon size is 1,280bp. Perform sequencing on the amplicon for confirmation.

Preventative measures

(1) Stamp out the culture pond for thorough disinfection and maintain good water quality. Strictly implement the 4 disinfection policies of fingerlings, feed, equipment and feeding site with the principles of qualifying, quantifying, timing and locating. Provide fresh feed and do not use deteriorated feed.

(2) Prior to stocking new fingerlings to the culture pond, immerse in 8~10mg/L bleach solution for 15~30 minutes.

(3) During outbreak seasons, disinfect the nearby areas of feeding site with bleach powder or quicklime, or apply 1mg/L bleach powder or 20~30mg/L quicklime to the whole pond for disinfection once every 15 days. If an outbreak occurs, these disinfectants can be applied daily for consecutive 3 days.

(4) For every one kilogram of fish, apply a total of 5g garlic or 0.02g allicin and 0.5g of sodium chloride in feed and treat daily for consecutive 3 days.

(5) For every one kilogram of fish, apply a grinded dry mixture with feed, composing 5g humifuse euphorbia herb, 5g common purslane, 5g Asian copperleaf or water pepper (sole or mixed) and 0.5g sodium chloride, treat daily for consecutive 3 days. If fresh ingredients are being used, then 25g humifuse euphorbia herb and common purslane, 20g Asian copperleaf or water peper should be applied.

(6) For every one kilogram of fish, apply a total of 20g dry green chiretta or 30g fresh green chiretta with 0.5g sodium chloride in feed and treat daily (twice a day) for consecutive 3 days.

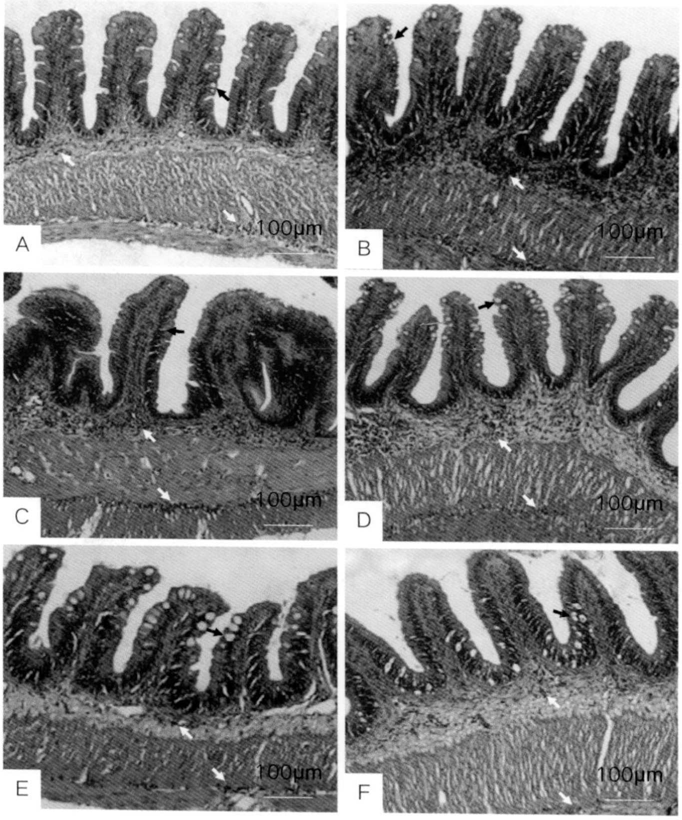

病鱼组织病变

白色箭头表示炎症细胞，黑色箭头表示杯状细胞

A. 正常草鱼肠道，黏膜下层以及环状肌层和纵向肌层之间的肌间隙炎症细胞轻度浸润

B. 感染1d后，黏膜下层炎症细胞大量浸润

C. 感染3d后，肠黏膜严重发炎，肠绒毛融合或脱落

D. 感染7d后，肠黏膜开始修复，杯状细胞数量明显增加

E. 感染14d后，黏膜下层炎症细胞数量明显减少

F. 感染21d后，肠黏膜病变几乎完全修复

[源自宋学宏]

Histological lesions of the intestines of the affected grass carp

In each panel. white arrows indicate inflammatory cells, and black arrows indicate goblet cells.

A. Intestines of healthy fish, showing mild infiltration of inflammatory cells in the submucosa and the intermuscular space between circular and longitudinal muscle layers.

B. Heavy infiltration of inflammatory cells in submucosa is shown 1 day after infection

C. 3 days after infection, severe inflammation in the intestinal mucosa and the fusion or shedding of intestinal villi were observed.

D. 7 days after infection, the intestinal mucosa began to repair and a marked increase in the number of goblet cells was observed

E. After 14 days, the number of inflammatory cells in the submucosa significantly decreased

F. After 21 days, intestinal mucosal lesions were almost completely repaired.

[Source: Xuehong Song]

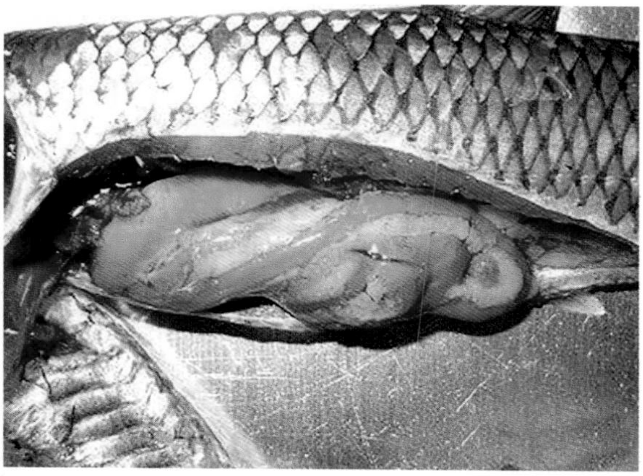

受感染的草鱼肠道

[源自利洋]

The intestines of the affected grass carp

[Source: LIYANG AQUATIC]

类结节病

疾病概述

【概述】 类结节病是一种发病率和死亡率都很高的传染性疾病。

【易感宿主】 主要感染养殖的黄尾鰤（*Seriola quinqueradiata*）幼鱼以及香鱼（*Plecoglossus altivelis*），养殖的黑棘鲷（*Acanthopagrus schlegeli*）幼鱼、赤点石斑鱼（*Epinephelus akaara*）、斑鳢（*Channa maculata*）、马面鲀（*Thamnaconus septentrionalis*）、条纹狼鲈（*Menticirrhus saxatilis*）、褐鳟（*Salmo trutta*）、大西洋鲑（*Salmo salar*）、真鲷（*Pagrosomus major*）、金头鲷（*Sparus aurata*）、牙鲆（*Paralichthys olivaceus*）、塞内加尔鳎（*Cynoglossus semilaevis*）、黄带拟鲹（*Pseudocaranx dentex*）、欧洲舌齿鲈（*Dicentrarchus labrax*）等，野生的美洲狼鲈（*Morone saxatilis*）、条纹狼鲈（*Menticirrhus saxatilis*）、大西洋油鲱（*Brevoortia tyrannus*）、鲻（*Mugil cephalus*）、红眼鱼（*Scardinius erythrophthalmus*）、天穹白鲑（*Coregonus zenithicus*）等。

【易感阶段】 主要感染养殖黄尾鰤幼鱼，2龄以上的大鱼也可被感染，死亡率可高达90%。

【发病水温】 感染发生于春末到夏季，发病最适水温为20 ～ 25℃。一般在温度25℃以上时很少发病，温度20℃以下不生病。秋季很少出现此病。

【地域分布】 广泛分布于欧美、日本、中国。

病原

（1）病原为美人鱼发光杆菌杀鱼亚种（*Photobacterium damselae* subsp. *piscicida*）。

（2）属弧菌科（Vibrionaceae）、发光杆菌属（*Photobacterium*）。

（3）革兰氏阴性菌，短杆状或球杆状，大小为（0.6 ～ 1.2）μm×（0.8 ～ 2.6）μm，不形成芽孢，兼性厌氧菌。

（4）不能在SS琼脂培养基和BTB琼脂培养基上生长。

（5）可在17 ～ 32℃生长，生长最适温度为20 ～ 30℃。生长的pH范围为6.8 ～ 8.8，最适pH为7.5 ～ 8.0。生长的盐度范围为5 ～ 30，最适盐度为20 ～ 30。

临床症状和病理学变化

（1）反应迟钝，食欲减退，离群独游或静止于网箱或池塘底部，继而不摄食。

（2）体色变黑，体表、鳍基、尾柄等处有不同程度充血，严重者全身肌肉充血。

（3）解剖可见肾、脾、肝、胰、心、鳔和肠系膜等组织或器官上有许多小白点，有的很微小，有的直径大至数毫米，多数为1mm左右，形状不规则，多数近于球形。

（4）血液中有许多细菌。肾脏白点数量多时，肾脏呈贫血状态。脾脏白点数量多时，脾脏肿胀而带暗红色。血液中菌落数量多时，在微血管内形成栓赛。

诊断方法

（1）细菌分离　使用含1%～2% NaCl的大豆胰蛋白胨培养基和脑心浸萃液态培养基、血平板或者海水培养基2216E，22℃培养2～4d，形成正圆形、无色、半透明、露滴状的菌落，有显著的黏稠性。

（2）生化鉴定　革兰氏阴性、无运动性，具有两极性的棒状杆菌，氧化酶和过氧化酶检测均为阳性，厌氧发酵，对O/129敏感，严格嗜盐。精氨酸二水解酶、脂酶和磷脂酶检测为阳性，吲哚、硝酸盐还原、脲酶、明胶酶、淀粉酶和产硫化氢检测为阴性。能分解葡萄糖、甘露糖、半乳糖和果糖产酸。

（3）16S rDNA鉴定

27F：5′-AGA-GTT-TGA-TCC-TGG-CTC-AG-3′。

1 492R：5′-GGT-TAC-CTT-GTT-ACG-ACT-T-3′，退火温度为52℃，扩增产物长度为1 500bp。

扩增产物经测序后判定。

防治方法

保证水源清洁，养殖期间应经常换用新水或保持流水，避免养殖水体富营养化，勿过量投饵或投喂腐败变质的生饵。

Pseudotuberculosis

Disease overview

[Disease Characteristic] An infectious bacterial disease with high morbidity and high mortality.

[Susceptible Host] Mainly affect aquaculture fish species such as juvenile of yellowtail (*Seriola quinqueradiata*), sweetfish (*Plecoglossus altivelis*), juvenile of blackhead seabream (*Acanthopagrus schlegeli*), Hong Kong grouper (*Epinephelus akaara*), blotched snakehead (*Channa maculata*), filefish (*Thamnaconus septentrionalis*), northern kingfish (*Menticirrhus saxatilis*), brown trout (*Salmo trutta*), Atlantic salmon (*Salmo salar*), red seabream (*Pagrosomus *major*). Gilt-head bream (*Sparus aurata*), olive flounder (*Paralichthys olivaceus*), Senegalese sole (*Solea senegalensis*), white trevally (*Pseudocaranx dentex*) and European bass (*Dicentrarchus labrax*). Also affect wild fish species such as white perch (*Morone Americana*), northern kingfish (*Menticirrhus saxatilis*), Atlantic menhaden (*Brevoortia tyrannus*), flathead grey mullet (*Mugil cephalus*), common rudd (*Scardinius erythrophthalmus*) and shortjaw cisco (*Coregonus zenithicus*).

[Susceptible Stage] Mainly affect juvenile of yellowtail, adult fish of over 2 year-old can also be infected. Mortality can reach 90%.

[Outbreak Water Temperature] Infection can occur from late spring to summer. The optimal outbreak water temperature is 20~25℃. In general, disease outbreaks seldom occur at a temperature higher than 25℃ or lower than 20℃. Disease outbreaks seldom occur in autumn.

[Geographical Distribution] Widely distributed in Europe, America, Japan and China.

Aetiological agent

(1) *Photobacterium damselae* subsp. *piscicida*

(2) Family: Vibrionaceae. Genus: *Photobacterium*.

(3) Gram-negative short rods or coccobacilli of (0.6~1.2)μm × (0.8~2.6)μm in size. Non-spore forming and facultative anaerobic bacteria.

(4) No growth on Salmonella-Shigella (SS) agar and Bromothymol Blue (BTB) agar.

(5) Able to grow from 17~32℃, with optimal temperature at 20~30℃. Can grow under the condition of pH 6.8~8.8, with optimal growing pH at 7.5~8.0, and salinity of 5~30, with optimal growing salinity at 20~30.

Clinical signs and pathological changes

(1) Lethargy, inappetence, segregated from group or sink to the bottom of the net cage or culture pond without consuming any feed.

(2) Darkened body color. Varying degree of hyperaemia on the body surface, fin base and tail. In severe case, diffuse muscular hyperaemia may be present.

(3) Autopsy reveals numerous whitish nodules in the kidneys, spleen, liver, pancreas, heart, swimming bladder and mesenteries. Most of the nodules are around 1mm in diameter, ranging from very small to a few mm in diameter, irregular in shape, mostly spherical.

(4) Numerous bacteria present in the blood. If numerous whitish nodules are present in the kidney, the kidney appears anaemic. If numerous whitish nodules are present in the spleen, the spleen appears swollen and dark red in color. If numerous bacterial colonies are present in the bloodstream, thrombi may be formed in the capillaries.

Diagnostic methods

(1) **Bacterial isolation** Bacterial culture using tryptic soy agar (TSA) plate, brain heart infusion (BHI) agar, blood agar or marine agar 2216E with 1%~2% sodium chloride, incubate at 22℃ for 2~4 days. The targeted colonies are uniformly round tear-shaped, colorless, translucent and apparently mucoid.

(2) **Biochemical identification** Gram-negative, non-motile, bipolar rod shape bacteria. Oxidase and peroxidase tests are negative. Anaerobic fermentation, sensitive to O/129 and obligative halophilic. L-arginine amidinohydrolase, lipase and phospholipase tests are positive. Indole, nitrate reduction, urease, gelatinase, amylase and H_2S production tests are positive. Fermenter for glucose, mannose, galactose and fructose.

(3) **16S rDNA identification** Use the primers 27F (5′-AGA-GTT-TGA-TCC-TGG-CTC-AG-3′) and 1,492R (5′-GGT-TAC-CTT-GTT-ACG-ACT-T-3′) to perform PCR amplification with annealing temperature at 52℃. The amplicon size is 1,500bp. Perform sequencing on the amplicon for confirmation.

Preventative measures and treatment

Keep the water source clean. During fish culture period, replace culture water with new water frequently or maintain constant water exchange to avoid eutrophication. Do not overfeed and maintain good quality of feed.

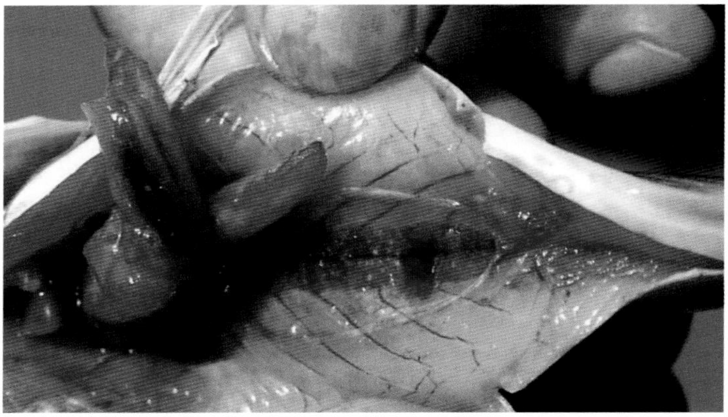

解剖检查时可见脾脏和肾脏出现小的白色斑点

[源自《新鱼病图鉴》，小川和夫]

Macroscopic of affected fish. Numerous pinpoint whitish lesions are present in the spleen and the kidney

[Source: *New Atlas of Fish Diseases*, Kazuo Ogawa]

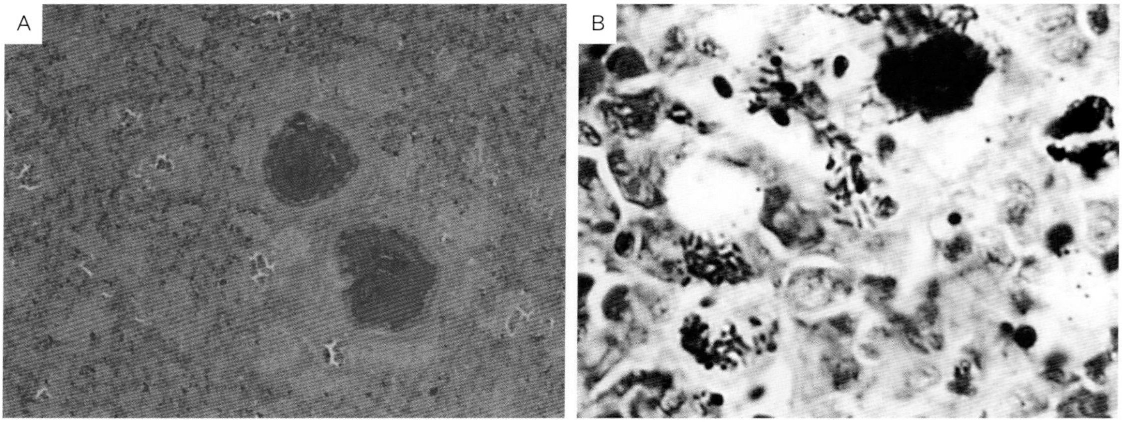

病鱼细胞及组织病变（HE染色）

A．菌团被炎细胞包围形成结节样结构

B．患部病原菌在其细胞内增殖

[源自《新鱼病图鉴》，小川和夫]

Histological lesions of affected fish (HE staining)

A．Multifocal necrotic foci with bacterial colonies are encircled by numerous inflammatory cells forming nodular lesions (HE staining)　B．Intracellular propagation of short-rod shaped bacteria

[Source: *New Atlas of Fish Diseases*, Kazuo Ogawa]

参考文献 References

陈世阳，李萱，1965. 鲤鱼疖疮病病原菌的分离与鉴定 [J]. 山东海洋学院学报，5(4): 134-137.

高桂生，张艳英，吉志新，等，2016. 半滑舌鳎致病性荧光假单胞菌的分离鉴定及其感染的病理损伤 [J]. 中国兽医学报，36(7): 1145-1150.

黄钧，施金谷，黄艳华，等，2012. 罗非鱼竖鳞病病原菌的分离鉴定及药敏试验 [J]. 南方农业学报，43(8): 1230-1234.

蒋依依，李安兴，2011. 诺卡菌特异性PCR快速检测方法的建立 [J]. 南方水产科学，7(6): 47-51.

可小丽，卢迈新，李庆勇，等，2013. 罗非鱼无乳链球菌鉴定及基于cfb和16S rRNA基因同源性分析 [J]. 中国农学通报，29(20): 52-62.

李莉萍，王瑞，黄婷，等，2013. 广东、海南、福建3省罗非鱼链球菌病流行菌株PCR鉴定和PFGE基因型分析 [J]. 西南农业学报，26(5): 2133-2140.

刘国文，2010. 草鱼细菌性烂鳃病病原菌的分离鉴定 [J]. 贵州农业科学，38(9): 134-135.

刘允坤，孙修勤，黄捷，等，2002. 牙鲆淋巴囊肿病的PCR诊断方法研究 [J]. 高技术通讯 (11): 87-89.

罗福广，黄杰，2016. 罗非鱼竖鳞病的诊断与防治 [J]. 科学养鱼 (3): 55-57.

彭新亮，吴海港，刘锦妮，等，2015. 金鱼竖鳞病病原菌的分离鉴定及药敏试验 [J]. 水产科技情报，42(5): 272-275.

彭亚，刘杰，胡大胜，等，2014. 黄沙鳖疖疮病的病原菌分离鉴定及药敏试验 [J]. 南方农业学报，45(7): 1296-1301.

王高学，白占涛，张向前，等，1999. 大鲵赤皮病病原分离鉴定及防治试验 [J]. 西北农业大学学报，27(4): 74-77.

王娜，张旻，景宏丽，等，2018. 斑点叉尾鮰病毒焦磷酸测序检测方法的建立 [J]. 畜牧与兽医，50(5): 89-92.

吴勇亮，苗鹏飞，于辉，等，2018. 鳜鱼致病性迟缓爱德华氏菌的分离鉴定及药敏试验 [J]. 南方农业学报，49(4): 794-799.

唐绍林，戚瑞荣，雷燕，2015. 鱼鳃霉病的诊断误区和临床诊断方法 [J]. 当代水产，40(2): 81-82.

解明旭，李华，叶仕根，等，2015. 鱼类鳃霉病的特点与诊治 [J]. 黑龙江畜牧兽医 (2): 129.

徐晔，段宏安，周毅，等，2013. 鱼立克次氏体病研究进展 [J]. 安徽农业科学，41(23): 9662-9666.

杨楠，张志强，吴同垒，等，2018. 半滑舌鳎源美人鱼发光杆菌美人鱼亚种的分离鉴定 [J]. 中国兽药杂志，52(2): 19-25.

于翔，2011. 鲤鱼竖鳞病病原的研究 [J]. 河北渔业 (7): 1-4.

袁向芬，石素婷，吕继洲，等，2018. 真鲷虹彩病毒LAMP检测方法的建立 [J]. 中国动物检疫，35(1): 95-99.

张德锋，李爱华，龚小宁，2014. 鲟分枝杆菌病及其病原研究 [J]. 水生生物学报，38(3): 495-504.

张海强，邵玲，2017. 鲤春病毒血症病毒核蛋白、磷蛋白与基质蛋白的表达、抗体制备及免疫原性比较 [J]. 水产学报，41(12): 1919-1927.

张旻，王姝，王娜，等，2017. 金鱼造血器官坏死病毒实时荧光定量PCR检测方法的建立[J]. 中国动物检疫，34(11): 99-103.

甄珍，2015. 山女鳟源致病性水霉菌的分离鉴定及其特性研究[D]. 哈尔滨：东北农业大学.

ADEL M. T., RENATE R., MICHELE T., 1997. Identification of mycobacteria infecting fish to the species level using polymerase chain reaction and restriction enzyme analysis [J]. Vet. Microbiol., 58 (2-4): 229-237.

AHNE W., 1986. Unterschiedliche biologische Eigenschaften 4 cyprinidenpathogener Rhabdovirus isolate [J]. J. Vet. Med. B, 33: 253-259.

AHNE W., BEARZOTTI M., BREMONT M., et al., 1998. Comparison of European systemic piscine and amphibian iridoviruses with epizootic haematopoietic necrosis virus and frog virus 3 [J]. J. Vet. Med. B, 45: 373-383.

AHNE W., BJöRKLUND H. V., ESSBAUER S., et al., 2002. Spring viraemia of carp (SVC) [J]. Dis. aquat. Org., 52: 261-272.

ALDRIN M., STORVIK B., FRIGESSI A., et al., 2010. A stochastic model for theassessment of the transmission pathways of heart and skeleton muscle inflammation, pancreas disease and infectioussalmon anaemia in marine fish farms in Norway [J]. Prev. Vet. Med., 93: 51-61.

ALMENDRAS F. E., FUENTEALBA I. C., JONES S. R. M., et al., 1997. Experimental infection and horizontal transmission of *Piscirickettsia salmonis* in freshwater-raised Atlantic salmon, *Salmo salar* L. [J]. J. Fish Dis., 20 (6): 409-418.

AMAGLIANT G., GIAMMARINI C., OMICCIODI E., 2007. Detection of *Listeria monocytogenes* using a commercial PCR kit and different DNA extraction methods [J]. Food Control, 18 (9): 1137-1142.

ANA I., MATA M., MAR B., et al., 2004. Development of a PCR assay for *Streptococcus iniae* based on the lactate oxidase (lctO) gene with potential diagnostic value [J]. Vet. Microbiol., 101 (2): 109-116.

ARAKAWA C.K., DEERING R.E., HIGMAN K.H., et al., 1990. Polymerase chain reaction (PCR) amplification of a nucleoprotein gene sequence of infectious haematopoietic necrosis virus [J]. Dis. Aquat. Org., 8: 165-170.

AUSTIN D. A., MCINTOSH D., AUSTIN B., 1989. Taxonomy of fish associated Aeromonas spp., with the description of Aeromonas salmonicida subsp. smithia subsp [J]. Syst. Appl. Microbiol., 11: 277-290.

BAEK Y., BOYLE J. A., 1996. Detection of channel catfish virus in adult channel catfish by a nested polymerase chain reaction [J]. J. Aquat. Anim. Health., 8 (2): 97-103.

BALDOCK F. C., BLAZER V. S., CALLINAN R. B., 2005. Outcomes of a short expert consultation on epizootic ulcerative syndrome (EUS): Re-examination of causal factors, case definition and nomenclature [M]. Diseases in Asian Aquaculture, V. Walker P., Laster R. & Bondad-Reantaso M. G., eds. Fish Health Section, Asian Fisheries Society, Manila, The Philippines, 555-585.

BEAKES G. W., GAY J. L., 1977. Gametangial nuclear division and fertilization in *Saprolegnia furcata* as observed by light and electron microscopy [J]. Trans. Br. mycol. Soc., 69 (3): 459-471.

BERGMANN S.M., SADOWSKI J., KIELPINSKI M., et al., 2010. Susceptibility of koi × crucian carp and koi × goldfish hybrids to koi herpesvirus (KHV) and the development of KHV disease (KHVD) [J]. J. Fish Dis., 33: 267-272.

BETTS A. M., STONE D. M., 2000. Nucleotide Sequence Analysis of the Entire Coding Regions of Virulent and Avirulent Strains of Viral Haemorrhagic Septicaemia Virus [J]. Virus Genes, 20: 259.

BONDAD-REANTASO M. G., LUMANLAN S. C., NATIVIDAD J. M., 1992. Environmental monitoring of the epizootic ulcerative syndrome (EUS) in fish from Munoz, Nueva Ecija in the Philippines [M]. Diseases in Asian Aquaculture I. Shariff M., Subasinghe R.P. & Arthur J.R., eds. Fish Health Section, Asian Fisheries Society, Manila. The Philippines, 475-490.

BOYLE J., BLACKWELL J., 1991. Use of polymerase chain reaction to detect latent channel catfish virus [J]. Am. J. Vet. Res., 52 (12): 1965-1968.

BYERS H. K., GUDKOVS N., CRANE M. S. J., 2002. PCR-based assays for the fish pathogen Aeromonas salmonicida I. Evaluation of three PCR primers sets for detection and identification [J]. Dis. Aquat. Organ., 49: 129-138.

CANO I., FERRO P., ALONSO M. C., et al., 2007. Development of molecular techniques for detection of lymphocystis disease virusin different ma-rine fish species [J]. J. Appl. Microbiol., 102 (1): 32-40.

CANO I., VALVERDE E. J., GARCIA-ROSADO E., et al., 2013. Transmission of lymphocystis disease virus to cultured gilthead seabream, *Sparus aurata* L., larvae [J]. J. Fish Dis., 36 (6): 569-576.

CHINCHAR V. G., HICK P., INCE I. A., et al., 2017. ICTV Virus Taxonomy Profile: Iridoviridae [J]. J. Gen. Virol., 98(5): 890-891.

CHOI S. K., KWON S. R., NAM Y. K., et al., 2006. Organ distribution of red sea bream iridovirus (RSIV) DNA in asymptomatic yearling and fingerling rock bream (*Oplegnathus fasciatus*) and effects of water temperature on transition of RSIV into acute phase [J]. Aquac., 256: 23-26.

CHOU H.Y., HSU C.C., PENG T.Y., 1998. Isolation and characterization of a pathogenic iridovirus from cultured grouper (*Epinephelus* sp.) in Taiwan [J]. Fish Pathol., 33: 201-206.

ÇIGDEM Ü., REMZIYE E. Y., *Capillaria* sp. Infestation and Bacterial Septicemia in the Angel Fish (*Pterophyllum scalare*) [J]. Journal of Fisheries Sciences, 7 (3): 232-240.

CORDERO H., CUESTA H., MESEGUER J., et al., 2016. Characterization of the gilthead seabream (*Sparus aurata* L.) immune response under a natural lymphocystis disease virus outbreak [J]. J. Fish Dis., 39 (12): 1467-1476.

COSTANZI C., 2005. Programming and Definition of Prevention Plans for Viral Haemorrhagic Septicaemia (VHS) and Infective Haematopoietic Necrosis (IHN) within the Territory of the Autonomous Province of Trento and State of Implementation [J]. Vet. Res. Commun., 29(Suppl 2): 147.

DíAZ-ROSALES P., ROMERO A., BALSEIRO P., et al., 2012. Microarray-Based Identification of Differentially Expressed Genes in Families of Turbot (*Scophthalmus maximus*) After Infection with Viral Haemorrhagic Septicaemia Virus (VHSV) [J]. Mar. Biotechnol., 14: 515.

DEOK C. L., HYUN J. H., SHIN Y. C., et al., 2012. Antibiograms and the estimation of epidemiological cut off values for *Vibrio ichthyoenteri* isolated from larval olive flounder, *Paralichthys olivaceus* [J]. Aquac., 342-343: (31-35).

DIAMANKA A., LOCH T. P., CIPRIANO R. C., FAISAL M., 2013. Polyphasic characterization of *Aeromonas salmonicida* isolates recovered from salmonid and non-salmonid fish [J]. J. Fish Dis., 36 (11): 949-963.

DIXON P. F., HILL B. J., 1983. Rapid detection of infectious pancreatic necrosis virus (IPNV) by the enzyme-linked immunosorbent assay (ELISA) [J]. Gen. Virol., 64 (2): 321-330.

DIXON P. F., HILL B. J., 1984. Rapid detection of fish rhabdoviruses by the enzyme-linked immunosorbent assay (ELISA) [J]. Aquac., 42: 1-12.

DO J. W., CHA S. J., KIM J. S., et al., 2005. Sequence variation in the gene encoding the major capsid protein of

Korean fish iridoviruses [J]. Arch. Virol., 150: 351-359.

DRURY S. E. N., GOUGH R. E., CALVERT I., 2002. Detection and isolation of an iridovirus from chameleons (*Chamaeleo quadricornis* and *Chamaeleo hoehnelli*) in the United Kingdom [J]. Vet. Rec., 150: 451-452.

FINSTAD Ø.W., FALK K., LøVOLL M., et al., 2012. Immunohistochemical detection of piscine reovirus (PRV) in hearts of Atlantic salmon coincide with the course of heart and skeletal muscle inflammation (HSMI) [J]. Vet. Res., 43: 27.

EMMENEGGER E. J., MEYERS T. R., BURTON T. O. et al., 2000. Genetic diversity and epidemiology of infectious haematopoietic necrosis virus in Alaska [J]. Dis. Aquat. Org., 40: 163-176.

ENGELKING H. M., HARRY J. B., LEONG J. C., 1991. Comparison of representative strains of infectious haematopoietic necrosis virus by serological neutralization and cross-protection assays [J]. Appl. Environ. Microbiol., 57: 1372-1378.

FAISAL M., DIAMANKA A., LOCH T. P., et al., 2016. Isolation and characterization of *Flavobacterium columnare* strains infecting fishes inhabiting the Laurentian Great Lakes basin [J]. J. Fish Dis., 40 (5): 637-648.

FOURRIER M., LESTER K., THOEN E., et al., 2014. Deletions in the highly polymorphic region (HPR) of infectious salmon anaemia virus HPRO haemagglutinin-esterase enhance viral fusion and influence the interaction with the fusion protein [J]. J. Gen. Virol., 95 (5): 1015-1024.

FRERICHS G. N., RODGER H. D., PERIC Z., 1996. Cell culture isolation of piscine neuropathy nodavirus from juvenile sea bass, *Dicentrarchus* labrax [J]. J. Gen. Virol., 77: 2067-2071.

GHIASI M., KHOSRAVI A. R., SOLTANI M., 2009. Characterization of *Saprolegnia* isolates from Persian sturgeon (*Acipencer persicus*) eggs based on physiological and molecular data [J]. Journal de Mycologie Médicale, 20 (1): 1-7.

GODAHEWA G.I., SEONGDO LEE, JEONGEUN KIM, et al., 2018. Analysis of complete genome and pathogenicity studies of the spring viremia of carp virus isolated from common carp (*Cyprinus carpio carpio*) and largemouth bass (*Micropterus salmoides*): An indication of SVC disease threat in Korea [J]. Virus Research, 255: 105-116.

GOMEZ D. K., SATO J., MUSHIAKE K., et al., 2004. PCR-based detection of betanodaviruses from cultured and wild marine fish with no clinical signs [J]. J. Fish Dis., 27: 603-608.

GOODWIN A. E., SADLER J., MERRY G. E., et al., 2009. Herpesviral haematopoietic necrosis virus (CyHV-2) infection: case studies from commercial goldfish farms [J]. J. Fish Dis., 32 (3): 271-278.

GRAHAM D. A., CURRAN W., ROWLEY H. M.; et al., 2002. Observation of virus particles in the spleen, kidney, gills and erythrocytes of Atlantic salmon, *Salmo salar* L., during a disease outbreak with high mortality [J]. J. Fish Dis., 25 (4): 227-234.

GRAY W. L., WILLIAMS R. J., GRIFFIN B. R., et al., 1999. Detection of channel catfish virus DNA in acutely infected channel catfish, *Ictalurus punctatus* (Rafinesque), using the polymerase chain reaction [J]. Journal of fish diseases, 22 (2): 111-116.

GRAZYELLA M. Y., ROBERTO C., FRANCISCO H. R., et al., 2019. Single-step genomic evaluation improves accuracy of breeding value predictions for resistance to infectious pancreatic necrosis virus in rainbow trout [J]. Genomics, 111 (2): 127-132.

GROTMOL S., NERLAND A. H., BIERING E., et al., 2000. Characterisation of the capsid protein gene from a nodavirus strain affecting the Atlantic halibut *Hippoglossus hippoglossus* and design an optimal reverse-transcriptase polymerase chain reaction (RT-PCR) detection assay [J]. J. Dis. Aquat. Org., 39: 79-88.

GROTMOL S., TOTLAND G. K., 2000. Surface disinfection of Atlantic halibut *Hippoglossus hippoglossus* eggs with ozonated sea-water inactivates nodavirus and increases survival of the larvae [J]. Dis. Aquat. Org., 39: 89-96.

GROVE S., FALLER R., SOLEIM K. B., et al., 2006. Absolute quantitation of RNA by a competitive real-time RT-PCR method using piscine nodavirus as a model [J]. J. Virol. Methods, 132: 104-112.

GUO M., SHI W., WANG Y., et al., 2018. Recombinant infectious haematopoietic necrosis virus expressing infectious pancreatic necrosis virus VP2 protein induces immunity against both pathogens [J]. Fish Shellfish Immunol., 78: 187-194.

HATAI K., EGUSA S., TAKAHASHI S., 1977. Study on the pathogenic fungus of mycotic granulomatosis - I. Isolation and pathogenicity of the fungus from cultured-ayu infected with the disease [J]. Fish Pathol., 12 (2), 129-133.

HE L., ZHANG A., PEI Y. et al., 2017. Differences in responses of grass carp to different types of grass carp reovirus (GCRV) and the mechanism of haemorrhage revealed by transcriptome sequencing [J]. BMC Genomics, 18: 452.

HEDRICK R.P., MCDOWELL T.S., AHNE W., 1992. Properties of three iridovirus-like agents associated with systemic infections of fish [J]. Dis. Aquat. Org., 13: 203-209.

HENGSTBERGER S.G., HYATT A.D., SPEARE R., et al., 1993. Comparison of epizootic haematopoietic necrosis and Bohle iridoviruses, recently isolated Australian iridoviruses [J]. Dis. Aquat. Org., 15: 93-107.

HESAMI S., ALLEN K. J., METCALF D., et al., 2008. Phenotypic and genotypic analysis of *Flavobacterium psychrophilum* isolates from Ontario salmonids with bacterial cold water disease [J]. Can. J. Microbiol., 54 (8): 619-629.

HILL B. J., UNDERWOOD B. O., SMALE C. J., et al., 1975. Physico-chemical and serological characterization of five rhabdoviruses infecting fish [J]. J. Gen. Virol., 27: 369-378.

HU Y., WANG H., ZHANG M., et al., 2009. Molecular analysis of the copper-responsive CopRSCD of a pathogenic *Pseudomonas fluorescens* strain [J]. J. Microbiol., 47 (3): 277-286.

ITANO T., KAWAKAMI H., KONO T., et al., 2006. Detection of fish no-cardiosis by loop-mediated isothermal amplication [J]. J. Appl. Microbiol., 100 (6): 1364-5072.

ITO T., KURITA J., MORI K., et al., 2016. Virulence of viral haemorrhagic septicaemia virus (VHSV) genotype Ⅲ in rainbow trout [J]. Vet Res., 47: 4.

JESSICA L. J., YUKIKO H., JEFFREY A. K., et al., 2012. Comparison of molecular detection methods for *Vibrio parahaemolyticus* and *Vibrio vulnificus* [J]. Food Microbiol., 30 (1): 105-111.

JOYCE J. E., PHILLIP H. K., CRAIG A. S., et al., 2004. Shoemaker Efficacy of *Streptococcus agalactiae* (group B) vaccine in tilapia (*Oreochromis niloticus*) by intraperitoneal and bath immersion administration [J]. Vaccine, 22 (27-28): 3769-3773.

JUAN J., BORREGO, ESTEFANIA J., et al., 2017. Lymphocystis disease virus: its importance in aquaculture [J]. Re. Aquac., 9 (2): 179-193.

KIANI S., SADAT N. N., NAZARI A., 2016. Detection of *Vibrio* species isolated from ornamental guppy fish in

Kashan, Isfahan, Iran fish culturing pounds [J]. Biological Journal of Microorganism, 4 (16): 43-45.

KIBENGE M. J., IWAMOTO T., WANG Y., et al., 2013. Whole-genome analysis of piscine reovirus (PRV) shows PRV represents a new genus in family *Reoviridae* and its genome segment S1 sequences group it into two separate sub-genotypes [J]. Virol. J., 10: 230.

KIELPINSKI M., KEMPTER J., PANICZ R., et al., 2010. Detection of KHV in freshwater mussels and crustaceans from ponds with KHV history in common carp (*Cyprinus carpio*) [J]. Israeli J. Aquaculture (Bamidgeh), 62: 28-37.

KIM H. J., RYU J. O., LEE S. Y., et al., 2015, Multiplex PCR for detection of the *Vibrio* genus and five pathogenic *Vibrio* species with primer sets designed using comparative genomics [J]. BMC Microbiology. 15: 239.

KIM D. H., HAN H. J., KIM S. M., et al., 2004. Bacterial enteritis and the development of the larval digestive tract in olive flounder, *Paralichthys olivaceus* (Temminck & Schlegel) [J]. J. fish dis., 27 (9): 497-505.

KOLODZIEJEK J., SCHACHNER O., DÜRRWALD R., et al., 2008. "Mid-G" region sequences of the glycoprotein gene of Austrian infectious haematopoietic necrosis virus isolates form two lineages within European isolates and are distinct from American and Asian lineages [J]. J. Clin. Microbiol., 46: 22-30.

KONGTORP R. T., TAKSDAL T., 2009. Studies with experimental transmission of heart and skeletal muscle inflammation in Atlantic salmon, *Salmo salar* L. [J]. J. Fish Dis., 32 (3): 253-262.

KVITT, H., HEINISCH, G., DIAMANT, A., 2008. Detection and phylogeny of Lymphocystivirus in sea bream *Sparus aurata* based on the DNA polymerase gene and major capsid protein sequences. Aquaculture 275, 58-63.

LAFRENTZ B., LAPATRA S., SHOEMAKER C., et al., 2012. Reproducible challenge model to investigate the virulence of *Flavobacterium columnare* genomovars in rainbow trout *Oncorhynchus mykiss* [J]. Dis. Aquat. Organ., 101: 115-122.

LEUNG K. Y., SIAME B. A., TENKINK B. J., et al., 2012. *Edwardsiella tarda*–Virulence mechanisms of an emerging gastroenteritis pathogen [J]. Microbes & Infection, 14 (1): 26-34.

LI L, LUO Y, YUAN J, et al., 2015. Molecular characterisation and prevalence of a new genotype of Cyprinid herpesvirus 2 in mainland China [J]. Can. J. Microbiol., 61 (6):381-387.

LIBO H., AIDI Z., YAPING W., et al., 2017. Differences in responses of grass carp to different types of grass carp reovirus (GCRV) and the mechanism of haemorrhage revealed by transcriptome sequencing [J]. BMC Genomics [serial online], 18:1-15.

LILLEY J. H., HART D., PANYAWACHIRA V., et al., 2003. Molecular characterization of the fish-pathogenic fungus *Aphanomyces invadans* [J]. J. Fish Dis, 26 (5): 263-275.

LING S. H., WANG X. H., LIM T. M., et al., 2001. Green fluorescent protein-tagged *Edwardsiella tarda* reveals portal of entry in fish [J]. FEMS Microbiol. Lett., 194 (2): 239-243.

LIU B., GONG Y., LI, Z. et al., 2016. Baculovirus-mediated GCRV *VP*7 and *VP*6 genes expression in silkworm and grass carp [J]. Mol Biol Rep., 43: 509.

LOCH T. P., FUJIMOTO M., WOODIGA S. A., et al., 2013. Diversity of fish - associated flavobacteria of Michigan [J]. J. Aquat. Anim. Health, 25: 149-164.

ZHENG L.P., GENG Y., YU Z.H., et al., 2018. First report of spring viremia of carp virus in *Percocypris* pingi in China [J]. Aquaculture, 493: 214-218.

MANRÍQUEZ R. A., VERA T., VILLALBA M. V., et al., 2017. Molecular characterization of infectious pancreatic necrosis virus strains isolated from the three types of salmonids farmed in Chile [J]. Virol. J., 14 (1): 17.

MARANCIK D. P., GREGORY D. W., 2013. A real-time polymerase chain reaction assay for identification and quantification of *Flavobacterium psychrophilum* and application to disease resistance studies in selectively bred rainbow trout *Oncorhynchus mykiss* [J]. J. Flood Risk Manag., 339 (2): 122-129.

MCCLEARY S. J., GILTRAP M., HENSHILWOOD K., et al., 2014. Detection of salmonid alphavirus RNA in Celtic and Irish Sea flatfish [J]. Dis. Aquat. Org., 109 (1): 1-7.

MCLOUGHLIN M. F., GRAHAM D. A., 2007. Alphavirus infections in salmonids–a review [J]. J. Fish Dis., 30: 511-531.

MCVICAR A. H., 1990. Infection as a primary cause of pancreas disease in farmed Atlantic salmon [J]. Bull. Eur. Assoc. Fish Pathol., 10 (3): 84-87.

MICHEL B., LEROY B., STALIN RAJ V., et al., 2010. The genome of cyprinid herpesvirus-3 encodes 40 proteins incorporated in mature virions [J]. J. Gen. Virol., 91: 452-462.

MIYOSHI Y., SUZUKI S. A., 2003. PCR method to detect *Nocardia seriolaein* fish samples [J]. Fish Pathol., 38 (3): 93-97.

MORSI G., HANEN S., HANEN N., et al., 2016. Molecular detection of the three major pathogenic *Vibrio* species from seafood products and sediments in Tunisia using Real-Time PCR [J]. J. Food Prot., 79 (12): 2086-2094.

NANJO A., SHIBATA T., SAITO M., et al., 2017. Susceptibility of isogeneic ginbuna *Carassius auratus langsdorfii* Temminck et Schlegel to cyprinid herpesvirus-2 (CyHV-2) as a model species [J]. J. Fish Dis., 40 (02): 157-168.

NORDSTROM J. L., VICKERY M. C. L., BLACKSTONE G. M., et al., 2007. Development of a multiplex real-time PCR assay with an internal amplification control for the detection of total and pathogenic *Vibrio parahaemolyticus* bacteria in oysters [J]. Appl. Environ. Microbiol., 73: 5840-5847.

NOVOTNY L., HALOUZKA R., MATLOVA L., et al., 2010. Morphology and distribution of granulomatous inflammation in freshwater ornamental fish infected with mycobacteria [J]. J. Fish Dis., 2010, 33 (12): 947-955.

NOVOTNY L., POKOROVA D., RESCHOVA S., et al., 2010. Firstclinically apparent koi herpesvirus infection in the Czech Republic [J]. Bull. Eur. Assoc. Fish Pathol., 30: 85-91.

OELCKERS K., VIKE S., DUESUND H., et al., 2014. Caligusrogercresseyi as a potential vector for transmission of Infectious Salmon Anaemia (ISA) virus in Chile [J]. Aquac., 420-421: 126-132.

OLIVARES-FUSTER O., BAKER J. L., TERHUNE J. S., 2007. Host - specific association between *Flavobacterium columnare* genomovars and fish species [J]. Syst. Appl. Microbiol., 30 (8): 624-633.

OSHIMA S., HATA J., SEGAWA C., et al., 1996. A method for direct DNA amplification of uncharacterized DNA viruses and for development of a viral polymerase chain reaction assay: Application to the red sea bream iridovirus [J]. Anal. Biochem., 242: 15-19.

OTTERLEI A., BREVIK Ø. J., JENSEN D., et al., 2016. Phenotypic and genetic characterization of *Piscirickettsia salmonis* from Chilean and Canadian salmonids [J]. BMC Vet. Res., 12: 55.

PARK S. B., AOKI T., JUNG T. S., 2012. Pathogenesis of and strategies for preventing *Edwardsiella tarda* infection in fish [J]. Vet. Res., 43 (1): 67.

PEñA B., ISLA A., HAUSSMANN D., et al., 2016. Immunostimulatory effect of salmon prolactin on expression of Toll-like receptors in *Oncorhynchus mykiss* infected with *Piscirickettsia salmonis* [J]. Fish Physiol. Biochem., 42: 509-516.

PEDUZZI R., BIZZOZERO S., 1977. Immunochemical investigation of four Saprolegnia species with parasitic activity in fish: Serological and kinetic characterization of a chymotrypsin-like activity [J]. Microb. Ecol., 3: 107.

PETTERSON E., SANDBERG M., SANTI N., 2009. Salmonid alphavirus associated with *Lepeoptheirus salmonis* (Copepoda: Caligidae) from Atlantic salmon, *Salmo salar* L. [J]. J. Fish Dis., 30: 511-531.

PLARRE H., DEVOLD M., SNOW M. et al., 2005. Prevalence of infectious salmon anaemia virus (ISAV) in wild salmonids in western Norway [J]. Dis. Aquat. Org., 66: 71-79.

RAND T.G., 1996. Fungal Diseases of Fish and Shellfish [M] //: Howard D.H., Miller J.D. Human and Animal Relationships. The Mycota (A Comprehensive Treatise on Fungi as Experimental Systems for Basic and Applied Research), 6, Berlin, Heidelberg: Springer.

SAHOO P. K., SWAMINATHAN T. R., ABRAHAM T. J., et al., 2016. Detection of goldfish haematopoietic necrosis herpes virus (Cyprinid herpesvirus-2) with multi-drug resistant *Aeromonas hydrophila* infection in goldfish: First evidence of any viral disease outbreak in ornamental freshwater aquaculture farms in India [J]. Acta Tropica, 161: 8-17.

SHAH S. Q. A., SØRUM H., 2014. Genetic localization of a TetR-like transcriptional regulator gene in *Pseudomonas fluorescens* isolated from farmed fish [J]. J. Appl. Genetics, 55: 541-544.

SHAO L., XIAO Y., HE Z. K., et al., 2016. An N-targeting real-time PCR strategy for the accurate detection of spring viraemia of carp virus [J]. Journal of Virological Methods, 229: 27-34.

SHIMAHARA Y., NAKAMURA A., NOMOTO R., et al., 2008. Geneticand phenotypic comparison of *Nocardia seriolae*, isolated from fish in Japan [J]. J. Fish Dis., 31 (7): 481-488.

SMIRNOV L. P., BOGDAN V. V., NEMOVA N. N., et al., 2000. Cytoplasmic protein vaccine against bacterial hemorrhagic septicemia (Aeromonosis) of fish [J]. Appl. Biochem. Microbiol., 36: 510-513.

SNOW M., MCKAY P., MCBEATH A. J., et al., 2006. Development, application and validation of a Taqman real-time RT-PCR assay for the detection of infectious salmon anaemia virus (ISAV) in Atlantic salmon (*Salmo salar*) [J]. Dev Biol., 126: 133-145.

STINGLEY R. L., GRIFFIN B. R., GRAY W. L., 2003. Channel catfish virus gene expression in experimentally infected channel catfish, *Ictalurus punctatus* (Rafinesque) [J]. J. fish dis., 26 (8): 487 - 493.

TAKAHASHI H., MIYA S., KIMURA B., et al., 2008. Difference of genotypic and phenotypic characteristics and pathogenicity potential of *Photobacterium damselae* subsp. *damselae*, between clinical and environmental isolates from Japan [J]. Microb. Pathog., 45 (2): 150-158.

TAKANO T., NAWATA A., SAKAI T., et al., 2016. Full-genome sequencing and confirmation of the causative agent of erythrocytic inclusion body syndrome in coho salmon identifies a new type of piscine Orthoreovirus [J]. Plos ONE, 11(10): 1-20.

TAKASHI A., IKUO H., KEN K., et al., 2007. Genome sequences of three Koi Herpesvirus isolates representing the expanding distribution of an emerging disease threatening koi and common carp worldwide [J]. J. Virol., 81 (10): 5058-5065.

TANAKA K., HEALTH I. B., 1984. The behaviour of kinetochore microtubules during meiosis in the fungus *Saprolegnia* [J]. Protoplasma, 120: 36-42.

TERCETI M. S., OGUT H., OSORIO C. R., 2016. *Photobacterium damselae* subsp. *damselae*, an emerging fish pathogen in the Black Sea: evidences of a multiclonal origin [J]. Appl. Environ. Microbiol., 82 (13).

TERCETI M. S., RIVAS A. J., ALVAREZ L., et al., 2017. Rst B regulates expression of the *Photobacterium damselae* subsp. *damselae* major virulence factors damselysin, phobalysin P and phobalysin C [J]. Front. Microbiol., 8: 582.

VANDERSEA M. W., LITAKER R. W., YONNISH B., et al., 2006. Molecular assays for detecting *Aphanomyces invadans* in ulcerative mycotic fish lesions [J]. Appl. Environ. Microbiol., 72 (2): 1551-1557.

VALDEBENITO S., AVENDANO-HERRERA R., 2009. Phenotypic, serological and genetic characterization of *Flavobacterium psychrophilum* strains isolated from salmonids in Chile [J]. J. Fish Dis., 32 (4): 321-333.

VILLALBA M., PéREA V., HERRERA L., et al., 2017. Infectious pancreatic necrosis virus infection of fish cell lines: Preliminary analysis of gene expressions related to extracellular matrix remodeling and immunity [J]. Vet. Immunol Immunopathol., 193-194: 10-17.

WIKLUND T., MADSEN L., BRUUN M. S., DALSGAARD I., 2000. Detection of Flavobacterium psychrophilum from fish tissue and water samples by PCR amplification [J]. J. Appl. Microbiol., 88 (2): 299-307.

XIAO L. K., JIAN G. W., ZE M. G., et al., 2009. Morphological and molecular phylogenetic analysis of two *Saprolegnia* sp. (Oomycetes) isolated from silver crucian carp and zebra fish [J]., Mycol. Res., 113 (5): 637-644.

YAN L., LIU H., Li X., et al., 2014. The VP2 protein of grass carp reovirus (GCRV) expressed in a baculovirus exhibits RNA polymerase activity [J]. Virol. Sin., 29: 86.

YU M. F., HUAI S. X., HONG S. W., et al., 2011. Outbreak of a cutaneous *Mycobacterium marinum* infection in Jiangsu Haian, China [J]. Diagn. Microbiol. Infect. Dis., 71 (3): 267-272.

ZHANG A., HE L., WANG Y., 2017. Prediction of GCRV virus-host protein interactome based on structural motif-domain interactions [J]. BMC Bioinformatics, 18: 145.

图书在版编目（CIP）数据

鱼类疾病诊断和防治图谱. 细菌、病毒卷 / 夏新生，薛汉宗主编. —北京：中国农业出版社，2022.3
ISBN 978-7-109-28652-8

Ⅰ. ①鱼… Ⅱ. ①夏… ②薛… Ⅲ. ①细菌性鱼病－鱼病防治－图谱②鱼类病毒病－鱼病防治－图谱 Ⅳ. ①S942－64

中国版本图书馆CIP数据核字(2021)第156753号

YULEI JIBING ZHENDUAN HE FANGZHI TUPU —— XIJUN BINGDU JUAN

中国农业出版社出版
地址：北京市朝阳区麦子店街18号楼
邮编：100125
责任编辑：王金环　郑　珂
版式设计：李　文　　责任校对：沙凯霖
印刷：北京中科印刷有限公司
版次：2022年3月第1版
印次：2022年3月北京第1次印刷
发行：新华书店北京发行所
开本：787mm × 1092mm　1/16
印张：11.75
字数：280千字
定价：148.00元